Digital Innovations in Architecture, Engineering and Construction

The Architecture, Engineering and Construction (AEC) industry is experiencing an unprecedented transformation from conventional labor-intensive activities to automation using innovative digital technologies and processes. This new paradigm also requires systemic changes focused on social, economic and sustainability aspects. Within the scope of Industry 4.0, digital technologies are a key factor in interconnecting information between the physical built environment and the digital virtual ecosystem. The most advanced virtual ecosystems allow to simulate the built to enable a real-time data-driven decision-making. This Book Series promotes and expedites the dissemination of recent research, advances, and applications in the field of digital innovations in the AEC industry. Topics of interest include but are not limited to:

- Industrialization: digital fabrication, modularization, cobotics, lean.
- Material innovations: bio-inspired, nano and recycled materials.
- Reality capture: computer vision, photogrammetry, laser scanning, drones.
- Extended reality: augmented, virtual and mixed reality.
- Sustainability and circular building economy.
- Interoperability: building/city information modeling.
- Interactive and adaptive architecture.
- Computational design: data-driven, generative and performance-based design.
- Simulation and analysis: digital twins, virtual cities.
- Data analytics: artificial intelligence, machine/deep learning.
- Health and safety: mobile and wearable devices, QR codes, RFID.
- Big data: GIS, IoT, sensors, cloud computing.
- Smart transactions, cybersecurity, gamification, blockchain.
- Quality and project management, business models, legal prospective.
- Risk and disaster management.

Francesco Carlucci · Ludovica Maria Campagna · Francesco Fiorito

Responsive Envelopes and Climate Change

State of the Art, Design Strategies, and Future Perspectives for Resilient Buildings

Springer

Francesco Carlucci
Department of Civil, Environmental, Land,
Building Engineering and Chemistry
(DICATECh)
Polytechnic University of Bari
Bari, Italy

Ludovica Maria Campagna
Department of Civil, Environmental, Land,
Building Engineering and Chemistry
(DICATECh)
Polytechnic University of Bari
Bari, Italy

Francesco Fiorito
Department of Civil, Environmental, Land,
Building Engineering and Chemistry
(DICATECh)
Polytechnic University of Bari
Bari, Italy

ISSN 2731-7269 ISSN 2731-7277 (electronic)
Digital Innovations in Architecture, Engineering and Construction
ISBN 978-3-031-58100-7 ISBN 978-3-031-58101-4 (eBook)
https://doi.org/10.1007/978-3-031-58101-4

This Springer imprint is published by the registered company Springer Nature Switzerland AG
The registered company address is: Gewerbestrasse 11, 6330 Cham, Switzerland

Preface

The growing frequency and magnitude of extreme weather events have increased the attention of the scientific and international community regarding climate change and the correlation between anthropized and external environment. Although climate change depends on several human activities, the strong incidence of building energy consumption on the global greenhouse gas emissions makes the building field particularly interested by this topic and sensitive to the mitigation actions.

The mitigation strategies in the building design field are mainly focused on two macro actions: the maximization of the energy efficiency of the building systems and the minimization of the energy losses. While the former refers to the building services design (HVAC, lighting, vertical transportation, renewable energy sources, etc.), the latter refers mainly to the envelope technologies. Focusing on the building envelope, the latest design strategies are shifting the attention from the maximization of the indoor-outdoor energy disconnection to a new approach that allows to react and adapt to indoor and/or outdoor environmental variations. Hence, the new role of the envelope is to adapt its properties—ranging from thermal and optical up to geometrical properties—according to different possible boundary conditions. It follows that, in this new perspective, the envelope is perceived as a responsive interface rather than a static shield. This new concept has started a fruitful scientific investigation in several fields, ranging from dynamic control of solar radiation up to the management of thermal inertia and heat fluxes, and currently several commercial products and prototypes are already available.

While these technologies have been widely tested on a yearly basis considering typical weather data, their responsive behaviour could have unexplored potential benefits in future scenarios considering the effects of climate change. The opportunity to adapt the envelope properties during different decades of the building service life, potentially increases the climate resilience of the built environment. This under-investigated research field is the key driver of the development of this book whose main aim is to provide all the technical background required to fully take advantage of the responsive behaviour of the building envelopes in current and future scenarios.

To pursue these goals, firstly, this book outlines a detailed description of climate change with a specific focus on its effects on the built environment. The reader will be provided with the theoretical underpinning of this phenomenon and all the technical concepts required to generate future weather data to conduct future-projected building energy analyses.

Secondly, this book provides a thorough review of the theoretical concepts underlying responsive technologies and a detailed description of the main currently—available responsive envelope technologies and their functioning. This preliminary theoretical framework is hence completed by a technical insight regarding the techniques required to model the responsive envelopes in the building performance simulation software.

Once the reader has been provided with the proper theoretical background, the authors test the potential and the limits of three different responsive technologies—based on completely different functioning criteria—on an office building case study. To fully explore the responsive behaviour of these systems, the simulations are conducted on both current and future scenarios in three different locations to investigate their climate-dependency. This application allows, on the one hand, to compare the behaviour of different devices in different contexts and, on the other hand, to highlight the key design inputs of a responsive component.

Finally, the authors provide the readers with an innovative perspective of future-projected energy analyses. In particular, according to the previous validated studies, this book reports an interesting approach to account for several future uncertainty sources—such as climate change, technological advancements, variations of the urban context, and building use—in the building energy assessments. Thanks to a new index—called Energy Flexibility Index—energy analyses conducted in several possible future scenarios can be easily summarized in a single value to allow reliable comparison between different technologies in future projections to find the most resilient technology for the designed building.

To sum up, overall, this book allows to explore the wide topics of responsive envelope design and climate change providing, on the one hand, all the required theoretical backgrounds and, on the other hand, practical tools and hints to implement these solutions in the design process. The book's purpose is to help researchers, architects, façade consultants, and students of building and architectural engineering to understand, simulate, and consider the emerging responsive envelope technologies as viable solutions to improve the energy consumption and the climate resilience of buildings in both current and future scenarios. Therefore, this book could be considered as a key text to bridge the gap between innovative technologies and design practice, providing all the required tools to begin to spread these systems improving the climate resilience of the built environment.

Bari, Italy

Francesco Carlucci
Ludovica Maria Campagna
Francesco Fiorito

Acknowledgements

Many of the concepts described and investigated in this book are the results of years of research and professional activities. It follows that the idea of this book has been shaped also by several mentors and colleagues that we have met during the last years; we are undoubtedly thankful to all of them for their fundamental contributions.

Moreover, we wish to acknowledge that this book has been developed, also thanks to the past contributions of Roel Loonen, Jan Hensen, and Kristoffer Negendahl. In particular, thanks to their availability and suggestions, Francesco Carlucci has had the opportunity to work in new inspiring environments—both in Eindhoven and Copenahagen—starting fruitful international collaborations that have led to two main milestones described in this book: the modelling of dynamic shading systems and the definition of the Energy Flexibility Index. Furthermore, this book would not be the same without the fruitful collaboration and exchange developed during the Open Glassroom project which has improved the quality of all the glass-related topics.

We are also thankful to Springer Nature and Pierpaolo Riva for the opportunity and support to publish this book. Finally, we are grateful to everyone who has supported us during the writing of this book and, in general, during the several challenges of academic life.

Authors' Contributions

This book is written by *Carlucci Francesco*, *Campagna Ludovica Maria,* and *Fiorito Francesco*. *Carlucci Francesco* (corresponding and main author) dealt with conceptualization, methodology, writing (original draft preparation), data curation, software, formal analysis, investigation, and visualization of the whole book. *Campagna Ludovica Maria* dealt with the conceptualization of the second chapter and with the writing of the first two chapters. *Fiorito Francesco* dealt with the review, editing, and supervision of the work.

In particular, the contribution of each author is reported for every chapter of the book as follows.

Chapter 1 "Introduction": *Carlucci Francesco, Campagna Ludovica Maria, Fiorito Francesco*

Chapter 2 "Climate Change and Future Challenges": *Carlucci Francesco, Campagna Ludovica Maria, Fiorito Francesco*

Chapter 3 "Responsive Envelope Technologies": *Carlucci Francesco*

Chapter 4 "Modelling Tools for Responsive Envelopes": *Carlucci Francesco, Fiorito Francesco*

Chapter 5 "Design Strategies": *Carlucci Francesco, Fiorito Francesco*

Chapter 6 "Evaluating the Building Resilience: *Energy Flexibility Index"*: *Carlucci Francesco.*

Contents

Abbreviations

Δh_m	Enthalpy of fusion per unit mass
μ	Mean
σ	Standard deviation
AF	Adaptive Facade
AGCM	Atmosphere Global Climate Model
AKE	Acclimated Kinetic Envelope
AOGCM	Coupled Atmosphere-Ocean General Circulation Models
API	Application Programming Interface
AR4	Fourth Assessment Report
AR5	Fifth Assessment Report
AR6	Sixth Assessment Report
ASHRAE	American Society of Heating, Refrigerating and Air-Conditioning Engineers
BLAST	Building Loads Analysis and System Thermodynamics
BPS	Building Performance Simulation
c	Specific heat
CABS	Climate Adaptive Building Shells
CBECS	Commercial Buildings Energy Consumption Survey
CDD	Cooling Degree Days
CEC	Conventional ElectroChromic
CH_4	Methane
c_L	Specific heat in the liquid phase
CMIP	Coupled Model Intercomparison Project
CO_2	Carbon dioxide
CO_2-eq	Equivalent carbon dioxide
ConFD	Conduction Finite Difference
COP	Coefficient of Performance
CS	Current Scenario
c_S	Specific heat in the solid phase
CTF	Conduction Transfer Function
CV	Coefficient of Variation

DBEC	Dual-Band ElectroChromic
DGU	Double Glazing Unit
DOE	Department of Energy
DSF	Dynamic Shading File
EC	Electrochromic
ECV	Energy Consumption Variation
EDSM	Equivalent Dynamic Shading Model
EER	Energy Efficiency Ratio
EFI	Energy Flexibility Index
EMS	Energy Management System
EPBD	Energy Performance of Buildings Directive
EPW	Energy Plus Weather
ERL	EnergyPlus Runtime Language
FAR	First Assessment Report
f_m	Fraction melted
GC	Gasochromic
GCM	Global Climate Model
G_f	Incident irradiance on the façade
GHG	Greenhouse Gas Emission
h	Enthalpy
HadCM3	Hadley Center Couple Model 3
HDD	Heating Degree Days
HS	Harmonized Scenarios
HVAC	Heating, Ventilation, and Air Conditioning
ICT	Information and Communication Technologies
idf	Input data file
IEA	International Energy Agency
IEQ	Indoor Environmental Quality
IPCC	Intergovernmental Panel on Climate Change
ITO	Tin doped indium oxide
LHS	Latent heat storage
LSPR	Localized surface plasmon resonance
m	Mass
MoO_3	Molybdenum oxide
N_2O	Nitrous oxide
NbO_x	Niobium oxide
NEC	Near Infrared Radiation switching EC
NECV	Normalized Energy Consumption Variation
NIR	Near Infrared Radiation
NOAA	National Oceanic and Atmospheric Administration
O_3	Ozone
OGCM	Ocean Global Climate Model
OS	Overshoot scenarios
PC	Photochromic
PCM	Phase Change Material

PDMS	Polydimethylsiloxane
PEC	PhotoElectroChromic
PTFE	Polytetrafluoroethylene
PV	Photovoltaic
PVC	PhotoVoltaChromic
Q_L	Latent heat
Q_s	Sensible heat
RBE	Responsive Building Element
RCM	Regional Climate Model
RCP	Representative Concentration Pathways
ref	Reference
SAR	Second Assessment Report
SDG	Sustainable Development Goal
SHGC	Solar Heat Gain Coefficient
$SHGC_b$	Solar Heat Gain Coefficient in bleached state
$SHGC_c$	Solar Heat Gain Coefficient in coloured state
SHS	Sensible Heat Storage
S_i	i-th future scenario
SMA	Shape Memory Alloy
SMH	Shape Memory Hybrids
SMP	Shape Memory Polymers
SRES	Special Report on Emissions Scenarios
SSM	Static Shading Model
SSP	Shared Socio-Economic Pathways
T	Temperature
t	Timestep
TAR	Third Assessment Report
TC	Thermochromic
T_{in}	Indoor air temperature
TMY	Typical Meteorological Year
T_o	Outdoor air temperature
TRY	Test Reference Year
U	Thermal transmittance
UNFCCC	United Nations Framework Convention on Climate Change
VIP	Vacuum Insulation Panel
VL	Visible Light
VLT	Visible Light Transmittance
VLT_b	Visible Light Transmittance in bleached state
VLT_c	Visible Light Transmittance in coloured state
VO_2	Vanadium dioxide
w	Weighting factor
WGII	Working Group II
WGIII	Working Group III
WO_3	Tungsten oxide
WWR	Windows to Wall Ratio

Chapter 1
Design Challenges in a Changing Environment

Abstract This chapter provides an overall introduction of the present book, describing an overview of the context in which the study is set. Firstly, the issue of climate change is introduced, identifying the complex relationship with the built environment, and in particular buildings, identified as responsible for the climate crisis, but also extremely vulnerable to its effects. Next, potential adaptation and mitigation measures involving buildings are explored, focusing on improving building envelopes. Accordingly, potential and challenges of responsive envelopes are introduced, highlighting the complexity in modelling, further increased by the need to consider the changing climate as an additional factor of variation.

Keywords Climate change · Built environment · Building performance simulations · Responsive envelopes · Architectural technology

The unusual and rapid variability that the Earth's climate system is experiencing is now undeniable, as well as its adverse effects are noticeable for all to see. The responsibility of human activities in this process is beyond doubt, so all efforts must be made to limit it, to avoid devastating consequences for the planet and for humans themselves. The involvement of the built environment among the driving factors of climate change has now been clarified, as well as its significant vulnerability to the unavoidable effects of climate variability. Consequently, its crucial role in international policies to reduce greenhouse gas emissions (GHGs)—in the ensuing efforts to limit climate change—is more relevant than ever. Likewise, the need to limit the built environment vulnerability by adopting measures to adapt to the changing climate is increasingly evident, requiring an overall rethink of the built spaces.

The adverse consequences of climate change are increasingly visible, becoming more and more pervasive, overcoming all expectations [1]. Not only natural systems, but also the built environment and human lives are vulnerable to these impacts, which affect every region of our planet, albeit unevenly [2]. Scientific evidence has proven that these fluctuations cannot be attributed to the natural forcings of the Climate System (such as earth orbit variations or solar variations), but are indisputably due to human activities, which cause an increase in GHGs emissions. Such increase leads

F. Carlucci et al., *Responsive Envelopes and Climate Change*, Digital Innovations in Architecture, Engineering and Construction,
https://doi.org/10.1007/978-3-031-58101-4_1

to an alteration of the natural greenhouse effect, trapping infrared radiation near the earth surface and thus leading to warming temperatures, as well as many other effects. Starting from the industrial revolution, the increase in GHGs has never stopped. For instance, a CO_2 concentration of 400 ppm was recorded in 2019, well beyond its natural range of variation, which has always been between 174 and 300 ppm, since 800 thousand years ago. Likewise, other GHGs like methane and nitrous oxide follow the same pattern of growth, dramatically affecting our Climate System.

The most tangible consequence of climate change is the increase in temperatures, involving both atmospheric, oceanic and land temperatures. In fact, a temperature increase of 1.1 °C with respect to pre-industrial levels (1850–1900) has been recorded for the period 2011–2020, with larger increases over land (1.59 °C) than over the ocean (0.88 °C).

Despite the efforts that may be made to limit the effects of climate change, temperatures are expected to keep rising: the best forecasts predict a 1.5 °C increase in the short term with respect to pre-industrial level, which will have significant impacts on the planet [2]. This will result in multiple risks, as well as the interaction of climate and non-climate risks, leading to issues that are harder and harder to manage. Indeed, climate change involves not only fluctuations in temperatures, but manifold effects, such as increase of globally precipitation over land, changes in near-surface ocean salinity, global retreat of glaciers, warming and acidification of the oceans, increase of global mean sea level, changes in the land biosphere. In addition, climate change exacerbates extreme adverse weather events, for instance increasing both frequency and magnitude of temperature anomalies.

It is worth noting that these effects do not homogeneously affect the regions of our planet, varying both between and withing regions. In fact, climate risks depend not only on climate hazards, but also on the exposure of systems and their vulnerability. Consequently, adopting proper adaptation measures to reduce both exposure and vulnerability of systems becomes critical, since—as mentioned—there are now irreversible effects to be addressed regardless of the efforts to tackle climate change. Clearly, adapting systems to the changing climate is not sufficient to avoid the impacts of climate change, but these must necessarily be mitigated by acting directly on the causes that lead to them.

Among the sectors that most contribute to climate change, the construction sector has now been widely recognized. Indeed, the Working Group III (WGIII) of the Intergovernmental Panel on Climate Change (IPCC) highlighted that 21% of the global GHG emissions registered in 2019 can be attributed to the building sector [3]. Referring only to CO_2 emissions—which constitutes the largest emissions among other GHGs—the construction sector accounted for 31% of all global emissions in 2019 [3], reaching 33% in 2022 [4, 5]. Such emissions derive from indirect emissions and direct emissions (related to building operations), as well as embodied emissions. Indirect emissions are associated with fuel combustion for energy production used as end use in buildings and constitute the largest component (18%). Direct emissions are those produced on-site related mainly to energy consumption (8%), while embodied emissions are associated with the production of building materials (7%) [4, 5].

Overall, building operations lead to 30% of final global energy consumption [4]. Clearly, the extent of consumption values differs by geographic areas, due to different incomes, energy prices, and climatic conditions. For instance, consumptions are mainly related to heating (both space heating and water heating) and space cooling, whose needs differ according to the climatic conditions of the building location [6]. For instance, in the European Union, buildings are responsible for 40% of energy consumption and 36% of direct and indirect gas emissions [7]. Likewise, in the USA the building sector accounted for nearly 40% of the total consumption in 2019 [8]. Different values—nearly 20%—can be found in China due to a very different income-energy price ratio [9].

Considering such alarming data, the international community has clearly targeted GHGs reductions to limit the effects of the current climate crisis. Consequently, over the past decades, numerous efforts have been made to establish specific goals to address such crisis, to be pursued jointly by the entire international community. Among the results of these efforts, two main documents are worth mentioning: the 2030 Agenda for Sustainable Development Goals [10] and the United Nation Paris Agreement Commitment [11]. While the Paris Agreement set out the goal of limiting the global temperature increase to 1.5 °C above pre-industrial levels, by proposing climate action plans, the 2030 Agenda establishes a global plan of action to address the world's most pressing challenges, outlining 17 Sustainable Development Goals (SDGs) among which there is a goal concerning climate action (SDG 13). Both agreements emphasized the central role of cities and communities in the fight against climate change. In fact, the Paris Agreement recognizes cities and national authorities as key actor in the process, while the SGDs dedicates an entire goal to sustainable cities and communities (SDG 11). At the building level, the implementation of building policies significantly relies on the adoption of regulatory policy instruments, which could involve regulatory instruments, information instruments and incentive instruments. For instance, European Union has developed a specific Regulatory Framework, the Energy Performance of Buildings Directive (EPBD), currently under revision as part of the Commission's latest "Fit for 55" work program, which has the ambitious goal of complete decarbonization of the building stock by 2050.

Currently, scientific evidence shows that nearly half of the energy demand of industrial and residential buildings is allocated to the heating, ventilation, and air conditioning (HVAC) system [12] and such HVAC consumption is strongly related to two main factors: on the one hand, the envelope losses [13], and on the other hand, the building heat gains [14]. The significant impact of HVAC consumption explains why energy efficiency efforts are focusing both on transportation and power generation, as well as on the improvement of the building envelopes [12, 15]. From a building consumption perspective, space heating, lighting, and cooling are the major energy end uses. Looking at their trend of variation in the last ten years (2020–2010), the space heating experienced a large decrease, accounting for a reduction of nearly 20%, followed by lighting consumption which showed a decrease of nearly 17%. Otherwise, cooling consumption registered a reduction of only 7% [16]. Although these variations are the result of a combination of many elements—among which climate change cannot be neglected—these trends can be explained considering two main facts: the spread of new highly efficient lighting technologies (LED), and the approach to energy conservation that has led to a significant increase in building insulation. While the former has clearly reduced lighting consumption, the latter has reduced heating demand, while increasing overheating in summer, especially in hot climates.

Until the early twenty-first century, the design of the building envelope was based on a conservative static approach aimed at minimizing thermal losses to the external environment. To that end, the disconnection between indoor and outdoor environment was maximized to ensure a good energy conservation in buildings [17]. Although this strategy succeeds in reducing thermal losses, it lacks consideration of the dynamism that characterizes the outdoor environment, in which climate variables—solar radiation, dry bulb air temperature, humidity, wind, rainfalls etc.—experience constant changes over the days. Indeed, the weather variables can fluctuate within a day (sub hourly and hourly variations), during days (daily and weekly variations), during seasons (monthly variations), and—considering the climate change—during decades. Moreover, if we consider that the users' behaviour is constantly changing, the boundary conditions of the building energy phenomena can be defined as extremely dynamic. Over time, this awareness led to a different conception of the building envelope, slowly introducing the new concept of responsive envelope in the scientific community.

In this case, the envelope is no longer conceived as a single behaviour static boundary, but rather as a multifunctional interface element [18], in which considering all interactions between the environment and buildings becomes essential. Thanks to this new approach, topics related to Indoor Environmental Quality (IEQ) can be addressed, overcoming the limitations of the traditional static approach in which issues such as overheating [19] or visual discomfort [20] could not be properly assessed. Furthermore, this shift in the conception of the building envelope allows for the evaluation of many other fundamental aspects of building physics in addition to energy consumption, such as natural ventilation [21], thermal and visual comfort [22, 23], as evidenced by the growing number of studies on IEQ in recent years [24].

Currently, the latest efforts in the field of envelope design research are focusing on the design of responsive technologies, which are systems capable of changing their properties (solar, thermal, optical, etc.) to reduce energy demand and improve indoor comfort. Clearly, the dynamism of the envelope leads to an increase in the complexity of the analyses, as it adds a further degree of freedom to the phenomenon. For instance, using dynamic energy analyses, not only outdoor weather conditions can change over time, but also the envelope properties can vary depending on other simulation parameters. This dependency is strictly related to the specific responsive technology as its behaviour can range considerably depending on its nature. For example, on the one hand, the intrinsically controlled systems spontaneously self-adjust their properties according to variations of the environmental parameters thanks to their constituent nature; while, on the other hand, the extrinsically controlled systems experience variations based on a sensor-processor-actuator architecture that allows the responsiveness. In addition, few technologies combine both extrinsically and intrinsically controls (i.e., thermoelectro-chromic windows). A further level of complexity comes from the uniqueness of each technology, which can operate on multiple factors. In fact, some strategies work on solar radiation control (kinetic shading, chromogenic glazing, etc.), others work on thermal inertia (phase change materials), others work on thermal transmittance (variable transmittance envelopes), as well as a combination of these aspects (such as the phase change material windows). However, despite the significant potential of these technologies, they still appear to be not widespread in the construction field. There are many reasons behind such slow spread, among which the lack of a common modelling and design knowledge appears to be a major one.

A large number of responsive technologies are available today, addressing several purpose-goals. Overall, the responsive technologies can act on a multiphysics domain, allowing an improvement of thermal comfort, indoor air quality, visual performance, acoustic comfort, or energy generation. Depending on the technologies, such domain can be addressed either simultaneously or singularly [25]. Each technology can be characterized by different response time—index of the system readiness—which ranges from seconds, to minutes, as well as hours or seasons. Regardless of the response time, the degree of response may also differ since the technologies may exhibit different degree in their variation (on/off or gradual). Likewise, the state change can occur at different scales of adaptation: from a molecular scale to a macroscopic scale. Moreover, as already mentioned, the variation of the systems can be triggered either by an external or internal stimulus, or by a combination of both. Clearly, the technologies can be applied at different spatial scale, involving building material, building components (façade element, wall, fenestration, roof) or the whole building. In addition, although these systems were conceived considering short and medium-term changes (from sub-hourly to seasonal variations), their responsiveness could be useful on broader time scales (decades) to address the aforementioned effects of climate change that could occur throughout the whole life cycle of the envelope. Therefore, the evaluation and optimization of these systems in both current and future weather scenarios could facilitate the deployment of these technologies, improving energy efficiency and user comfort in a broader perspective.

Despite the envelope responsiveness is a very multi-faceted topic, three main clusters based on their functioning are identified in this book to ease the comprehension: the improvement of the thermal inertia, the control of the solar radiation, and the control of the heat flux. Within each class, this book explores the main responsive technologies and analyses their technological core, their strengths and weaknesses, their modelling technique, and their design. For example, among the systems that allows for an improvement of the thermal inertia, the Phase Change Materials (PCMs) are considered applied to both opaque and transparent envelopes (even if the PCM windows are still prototypes [26, 27]). Referring to the control of the solar radiation, the switchable windows and the dynamic shadings appear to be the most promising responsive technologies which are considered and detailed in this book. The former responds at a microscopic level, showing molecular changes due to the interaction with solar radiation, the latter works at a macroscopic level, stopping solar radiation through their movements. Thanks to their capability to maximize the solar gains during winter and reduce the solar loads during summer significant improvements can be achieved in the building performance. Electrochromic windows, thermochromic windows, photochromic, photoelectrochromic and photovoltachromic glazing, Shape Memory Alloys dynamic shadings, mechanically controlled kinetic systems are only few of the systems based on the control of the solar radiation described in this book. Finally, systems that act on heat flow control, try to find a balance between winter and summer needs, minimizing the thermal transmittance when required (i.e., during winter) and increasing the heat flows when needed (i.e., during summer when the systems are off and the outdoor temperature is lower than the indoor temperature).

As already mentioned, although the responsive technologies are versatile and promising, the simulation of their behaviour still appears to be challenging and requires specific computational skills which are still not largely common. However, not each responsive system can be easily simulated due to the intrinsic complexity of the phenomena, inhibiting prediction of their behaviour. In fact, unlike the static envelopes, the responsive systems are much more dependent on the context in which they operate, since factors like latitude, altitude, longitude, weather, users, control strategy of activation can significantly affect their performance. Consequently, the degree of freedom of responsive models are higher than those typical of traditional buildings, involving not only the outdoor and indoor environments, but also the envelope properties, affecting each other.

Currently, energy modelling software incorporates only a few predefined, simple, and easily accessible responsive technologies—like dynamic curtains—while more sophisticated systems require specific programming knowledge and/or workarounds, thus being more time consuming and less accessible. Thus, while many studies are conducting technology analysis on these devices, their energy analysis is lagging and still not widespread. In addition, energy analyses outcomes are closely related to the climate, settings and model considered, adding further complexity to a topic still under development. Consequently, there are few energy studies available on the same technology that are also difficult to compare. All these elements are slowing the diffusion of responsive systems, thus constituting a missed opportunity to innovate the building envelope sector with devices that enable it to accommodate environmental changes rather than merely provide a disconnection between the indoor and outdoor environment.

This complex topic is further difficult due to the uncertainty associated with climate change, which results in variation in external conditions that must be considered, as it will be inevitable. Climate change assessment adds additional degrees of uncertainty to the evaluation of these phenomena as they can be considered in different ways in building-scale applications. In fact, climate projections—which are typically the domain of climate science—cannot be directly used in building-scale simulations but must necessarily be scaled to be used within building simulation software, as these require hourly weather files. Indeed, climate projections are generated by global climate models (GCMs), first validated against past weather observations, and then forced by future emissions scenarios—associated with different potential world developments—to simulate future weather conditions. The GCMs output are available in terms of predicted variations in climate variables with respect to a reference period. However, such outputs have inadequate spatial and temporal resolution for direct application in building performance assessments, which require hourly weather data with a local spatial resolution. Hence, GCMs outputs must be spatially and temporally downscaled. To this end, different methodologies have been developed, involving statistical and dynamical downscaling approaches. Likewise, some tools have been developed to easily create future climate files suitable to be directly used in the buildings assessments.

This brief introduction on the two main topics of the book underlines the complex framework that characterizes the analyses of responsive devices in future scenarios. Moreover, additional uncertainties related to future developments (technological advancements, change of building use, urban changes, etc.) increase the unpredictability of the behaviour of our built environment in future scenarios. Clearly, a huge number of possible scenarios can be identified; nevertheless, increasing the number of possible scenarios means more possible boundary conditions where the buildings could operate, with a consequent potential great difference between the design assumptions and the actual operating context. Considering different options is undoubtedly an advisable approach but it leads to a number of results that is often hardly to manage and compare. To that end, once all the theoretical and practical backgrounds are defined, this book also proposes an innovative approach to manage this multitude of scenarios and technologies in a single comprehensive index.

To summarise, multiple questions are still open in the research community with reference to adaptive technologies, and new scenarios are opening up in thinking about exploiting their responsiveness not only in the short and medium term but also in the long term, to cope with the effects of future changes on buildings.

In this book, the authors explore such issues in depth, with the aim of providing a description of the most promising adaptive technologies, testing their critical issues and strengths in design under different climate scenarios.

The following part of the book is divided into 5 chapters, each one with a specific focus and purpose. In detail, Chaps. 2, 3 and 4 provide an overview of the state of the art concerning all the above-mentioned topics, while Chaps. 5 and 6 are more focused on design applications.

Chapter 2 provides an in-depth exploration of the topic of climate change, first presenting the basics behind climate change theory as well as the complex relationship with the built environment. Finally, existing methodologies for modelling climate change in the context of building research are explored, dwelling on the tools available for generating future projections that can be directly used in building simulations.

Following this, Chap. 3 firstly explores the current nomenclature and classification systems referred to responsive technologies, to allow a clear understanding of the proposed concepts. Then, the different functioning criteria are grouped and described analyzing each specific technology in detail.

Chapter 4 further explores the topic of responsive systems giving an overview of the modelling tools available. This section provides the required knowledge to model the responsiveness of a device focusing not only on standard modelling approaches—based on existing Energy Plus modules—but also on more complex advanced modelling techniques.

These theoretical concepts are then applied on real energy simulations in Chap. 5 where three different responsive approaches on building envelopes are tested on a reference case study. The same reference model is simulated in different locations and climate scenarios to bring out a new awareness of possible weaknesses and strengths of responsive technologies, to properly understand their functioning and maximize their efficiency.

Finally, Chap. 6 proposes a new way to consider the great influence of different future uncertainty sources—ranging from climate change to technology advancements—on both responsive or static technologies in the building energy assessment. Indeed, the proposed Energy Flexibility Index allows to summarize the behaviour of a technology in different scenarios (environmental, urban, technological, etc.) in a single index simplifying the assessment of the building resilience in a changing environment.

References

1. Intergovernmental Panel on Climate Change (IPCC), *Climate Change 2021: The Physical Science Basis. Contribution of Working Group I to the Sixth Assessment Report of the Intergovernmental Panel on Climate Change* (Cambridge University Press, Cambridge, United Kingdom and New York, NY, USA, 2023). ISBN 9781009157896
2. Intergovernmental Panel on Climate Change (IPCC), *Climate Change 2022—Impacts, Adaptation and Vulnerability* (Cambridge University Press, 2023), ISBN 9781009325844
3. Intergovernmental Panel on Climate Change (IPCC), *Climate Change 2022: Mitigation of Climate Change. Contribution of Working Group III to the Sixth Assessment Report of the Intergovernmental Panel on Climate Change* (Cambridge University Press, Cambridge, UK and New York, NY, USA, 2022)
4. International Energy Agency (IEA), *Tracking Clean Energy Progress 2023*, Paris (2023)
5. IEA; IRENA; UN Climate Change High Level Champions Breakthrough Agenda Report 2023, Paris (2023)
6. International Energy Agency (IEA). *World Energy Outlook 2023* (2023)
7. European Commission (EU). *In Focus: Energy Efficiency in Buildings* (2020)
8. U.S. Energy Information Administration. August 2020: Monthly Energy Review (2020)
9. X. Cao, X. Dai, J. Liu, Building energy-consumption status worldwide and the state-of-the-art technologies for zero-energy buildings during the past decade. Energy Build **128**, 198–213 (2016). https://doi.org/10.1016/j.enbuild.2016.06.089
10. UN General Assembly. *Transforming Our World: The 2030 Agenda for Sustainable Development* (2015)
11. United Nations. *Paris Agreement to the United Nations Framework Convention on Climate Change.*
12. K.J. Chua, S.K. Chou, W.M. Yang, J. Yan, Achieving better energy-efficient air conditioning—a review of technologies and strategies. Appl. Energy **104**, 87–104 (2013). https://doi.org/10.1016/j.apenergy.2012.10.037

13. L.C. Ng, A.K. Persily, S.J. Emmerich, Consideration of envelope airtightness in modelling commercial building energy consumption. Int. J. Vent. **12**, 369–378 (2014). https://doi.org/10.1080/14733315.2014.11684030
14. R. Elsland, I. Peksen, M. Wietschel, Are internal heat gains underestimated in thermal performance evaluation of buildings? Energy Proc **62**, 32–41 (2014). https://doi.org/10.1016/j.egypro.2014.12.364
15. W. Feng, Q. Zhang, H. Ji, R. Wang, N. Zhou, Q. Ye, B. Hao, Y. Li, D. Luo, S.S.Y. Lau, A review of net zero energy buildings in hot and humid climates: experience learned from 34 case study buildings. Renew. Sustain. Energy Rev. **114**, 109303 (2019). https://doi.org/10.1016/j.rser.2019.109303
16. United Nations Environment Programme. *Global Status Report for Buildings and Construction: Towards a Zero-Emission, Efficient and Resilient Buildings and Construction Sector* (2021)
17. R.C.G.M. Loonen, *Approaches for Computational Performance Optimization of Innovative Adaptive Façade Concepts*. Technische Universiteit Eindhoven (2018)
18. M. Perino, V. Serra, Switching from static to adaptable and dynamic building envelopes: a paradigm shift for the energy efficiency in buildings. J. Facade Des. Eng. **3**, 143–163 (2015). https://doi.org/10.3233/FDE-150039
19. R.S. McLeod, C.J. Hopfe, A. Kwan, An investigation into future performance and overheating risks in Passivhaus dwellings. Build. Environ. **70**, 189–209 (2013). https://doi.org/10.1016/j.buildenv.2013.08.024
20. M.A. Fasi, I.M. Budaiwi, Energy performance of windows in office buildings considering daylight integration and visual comfort in hot climates. Energy Build **108**, 307–316 (2015). https://doi.org/10.1016/j.enbuild.2015.09.024
21. A. Aflaki, N. Mahyuddin, Z. Al-Cheikh Mahmoud, M.R.A. Baharum, Review on natural ventilation applications through building façade components and ventilation openings in tropical climates. Energy Build **101**, 153–162 (2015). https://doi.org/10.1016/j.enbuild.2015.04.033
22. A. Cannavale, F. Fiorito, D. Resta, G. Gigli, Visual comfort assessment of smart photovoltachromic windows. Energy Build **65**, 137–145 (2013). https://doi.org/10.1016/j.enbuild.2013.06.019
23. S.M. Hosseini, M. Mohammadi, A. Rosemann, T. Schröder, J. Lichtenberg, A morphological approach for kinetic façade design process to improve visual and thermal comfort: review. Build. Environ. **153**, 186–204 (2019). https://doi.org/10.1016/j.buildenv.2019.02.040
24. Y. Alhorr, M. Arif, M. Katafygiotou, A. Mazroei, A. Kaushik, E. Elsarrag, Impact of indoor environmental quality on occupant well-being and comfort: a review of the literature. Int. J. Sustain. Built Environ. **5**, 1–11 (2016). https://doi.org/10.1016/j.ijsbe.2016.03.006
25. R. Loonen, J. Rico-Martinez, F. Favoino, M. Brzezicki, C. Menezo, G. La Ferla, L. Aelenei, Design for façade adaptability: towards a unified and systematic characterization, in *10th Conference on Advanced Building Skins* (2015)
26. F. Goia, M. Perino, V. Serra, Experimental analysis of the energy performance of a full-scale PCM glazing prototype. Sol. Energy **100**, 217–233 (2014). https://doi.org/10.1016/j.solener.2013.12.002
27. L. Giovannini, F. Goia, V.R.M. Lo Verso, V. Serra, Phase change materials in glazing: implications on light distribution and visual comfort. Preliminary results. Energy Proc. **111**, 357–366 (2017). https://doi.org/10.1016/j.egypro.2017.03.197

Chapter 2
Climate Change and Future Challenges

Abstract This chapter aims to present the topic of climate change, discussing the future challenges in addressing the climate crisis. The document presents the phenomenon from a theoretical perspective, explaining how the climate system works and identifying the main causes and effects of climate variations. The concepts of vulnerability, adaptability and mitigation to climate change are also introduced. Next, the chapter explores the complex relationship between climate crisis and the built environment, focusing on buildings. Finally, the use of future climate projections in building research field is deepened, presenting the principles of climate modelling and the main tools useful to generate future climate files.

Keywords Climate change · Built environment management · Climate modelling · Climate projections · Building performance simulations · Future climate files

The introductory section has highlighted that climate change is now at the core of all international debates and represents a priority to be considered in the development policies of the planet, included those related with the built environment. In this framework, understanding the Earth's Climate System functioning, its variations over time, and human influence are key aspects for appreciating the potential impacts on human life and the strategies needed to counter it.

The present chapter aims to present the basis of the theory behind climate change as well as its challenging relationship with the built environment. Given the magnitude of the topic, the chapter is divided into three sections. The first section deals with climate science, providing an overview of the climate emergency and identifying the main causes and effects starting from the Intergovernmental Panel on Climate Change (IPCC) outcomes. The second section highlights the complex relationship between climate change and the built environment, with a focus on the relationship with buildings. The third section aims to provide an in-depth analysis on climate projections, first providing a theoretical background about their generation within the field of climate modelling and then analysing the tools adopted in the field of built environment research.

F. Carlucci et al., *Responsive Envelopes and Climate Change*, Digital Innovations in Architecture, Engineering and Construction,
https://doi.org/10.1007/978-3-031-58101-4_2

2.1 Causes and Effects of a Climate Emergency

Two main definitions can be adopted to describe what climate change is.

The former, provided by the United Nations Framework Convention on Climate Change (UNFCCC), defines climate change as "*a change of climate which is attributed directly or indirectly to human activity that alters the composition of the global atmosphere and which is in addition to natural climate variability observed over comparable time periods*" [1]. The latter, provided by the Intergovernmental Panel on Climate Change (IPCC), defines climate change as "*a change in the state of the climate that can be identified (e.g., by using statistical tests) by changes in the mean and/or the variability of its properties and that persists for an extended period, typically decades or longer*" [2]. Accordingly, while the UNFCCC binds climate change to purely anthropogenic causes, the IPCC adopts the term more broadly, whether it results from natural or human causes.

The present section aims to provide a general overview on the topic of climate change, starting with the physical science behind the climate change theory. To that end, a brief explanation of the Global Climate System is presented (Sect. 2.1.1), highlighting its variability characteristics (Sect. 2.1.2). Next, scientific evidence of human responsibility for the unusual variation in the Climate System is reported, deepening the causes of the increase in human-caused CO_2 emissions (Sect. 2.1.3). Finally, impacts, vulnerability, and adaptability are investigated (Sect. 2.1.4), along with mitigation potential (Sect. 2.1.5).

The overall section is drawn from the main outcomes of the IPCC Sixth Assessment Report, which represents its core Refs. [2–4].

2.1.1 The Global Climate System

The Global Climate System is made up of five major interacting components, described hereafter:

- the atmosphere, mainly composed of nitrogen (78.1%), oxygen (20.9%), and argon (0.93%). In addition, the 0.1% of the atmosphere is composed of the so-called Greenhouse Gases (GHGs), namely water vapour (H_2O), carbon dioxide (CO_2), methane (CH_4), nitrous oxide (N_2O) and ozone (O_3).
- the hydrosphere, involving all liquid surface and subterranean water.
- the cryosphere, involving continental glaciers, sea ice and permafrost, as well as the ice sheets of Greenland and Antarctica.
- the lithosphere, involving vegetation and soils at the land surface.
- the biosphere.

The Climate System can be affected by different external forcing mechanisms. However, the major energy source that drives the whole system is the solar radiation [5]. Like all the existing objects, the Sun emits electromagnetic radiation, some of which reaches our planet. Each square meter of Earth's surface—considered externally to the atmosphere—is reached by an average of 342 Watts of solar radiations each year. Such radiation is distributed as follows. About one-third of the incoming solar radiation (approx. 107 W/m^2) is immediately reflected back into the space, by clouds and atmosphere ($\approx$77 W/m^2), as well as by earth surface ($\approx$30 W/m^2). About one-fifth of the incoming solar radiation (approx. 67 W/m^2) is absorbed by the atmosphere. About one-half of the incoming solar radiation (approx. 168 W/m^2) is absorbed by the Earth's surface (both oceans and land), which warms up. Then, the Earth's surface returns such heat, mainly in the form of infrared radiation (but also of sensible heat and water vapor). While some of this radiation returns to space, roughly 70% is absorbed by GHGs found in the atmosphere, which are able to absorb and re-emit infrared radiation in all directions, including towards the Earth's surface, leading to an increase of temperature near the Earth's surface. This process, known as the natural greenhouse effect, makes it possible to keep an average temperature on Earth suitable for life.

Under steady state conditions, a balance is required between incoming solar radiation and the outgoing radiation. However, such energy balance can be perturbed, leading to variations in the Climate System.

2.1.2 Climate System Variability

Over time, the Climate System may experience fluctuations from typical averages, known as climate variability (see Fig. 2.1). On the one hand, such variations may be intrinsic, that is, due to changes in the internal dynamics of the System (internal variability). On the other hand, the variations can be extrinsic (forced variability), that is, due to changes in the external forcings of the System, natural or anthropogenic. Among the natural forcings of the Climate System, there are solar variations, volcanic eruptions, as well as—on longer time scales—orbital forcing and plate tectonics. Otherwise, among anthropogenic forcings, researchers find out the changing in the atmospheric composition (increase of GHGs and aerosol concentrations from burning fossil fuels) and changing in the land-use, which affect the greenhouse effect explained in the previous section.

There is no doubt that the Climate System experienced many naturally variations in the past. For instance, thanks to paleoclimatic archives, climatic conditions before the instrumental era have been reconstructed, highlighting the existence of eight complete glacial-interglacial cycles over the past 800 thousand years. During these cycles, atmospheric CO_2 concentrations, global mean surface temperature and global mean sea level—considered as the key indicators of climate change—experienced natural variations recurring in dominant cycles of about 100 thousand years. In that case, the main cause was natural and relied upon the oscillations in the Earth's orbit and consequent feedbacks on multi-millennial time scales [2].

Therefore, if climate change is a natural phenomenon that our planet has already experienced in the past, what are scientists concerned about? The main issue is the

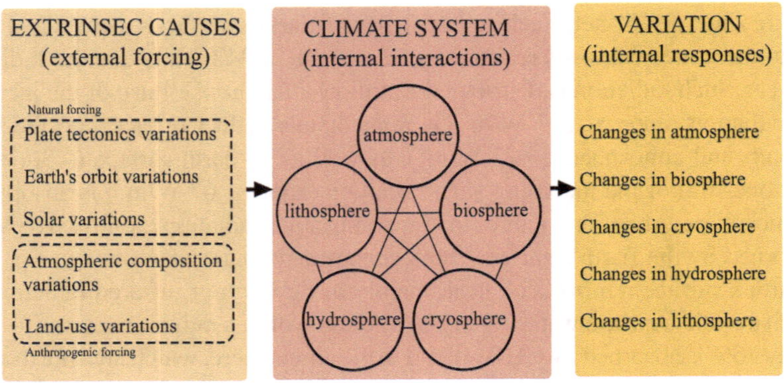

Fig. 2.1 Climate system variability

fast rate and great magnitude at which the Climate System is changing, as many of the climatic variations currently observed are unprecedented in thousands of years. For instance, over the past 800 thousand years, atmospheric CO_2 concentration always remained within the range between 174 and 300 ppm, thus considered as a "natural" global-scale range of variation. However, starting from 1750, the growth in CO_2 concentration has far exceeded such natural variations, reaching 410 ppm in 2019. Likewise, other GHGs like CH_4 and N_2O followed the same pattern of increase, dramatically affecting our Climate System. Such rapid and extensive changes cannot be explained by natural drivers: the observed growths have been unequivocally attributed to human activity.

2.1.3 Human-caused Climate Change

The influence of human activities on the observed changes in the climate system is now undisputed and has led to the recognition of a new geological epoch called the Anthropocene, started in the mid-twentieth century. The name comes from the awareness that, in such era, the major components of the Climate System are altered by human activities, resulting in effects that will remain in the permanent geological record. Indeed, natural causes (like solar variations or volcanic activity) have now been shown to have a negligible impact on the climate system. Otherwise, human activities have unequivocally increased GHGs concentration. As explained in the previous section, GHGs in the atmosphere can alter the natural greenhouse effect, trap infrared radiation near the surface and thus leading to warming temperatures, and many other effects. Consequently, "human influence has warmed the atmosphere, the ocean and the land", leading to atmosphere, ocean, cryosphere, and biosphere changes. Carbon Dioxide appears to be the major driver of global warming, followed by methane, nitrous oxide, and chlorofluorocarbons. It is worth noting that, the latter do not exist in nature, but they are industrial products.

According to the Sixth IPCC report, each of the last four decades has been successively warmer than any decade that preceded it since 1850. Global surface temperature in the period 2001–2020 was 0.99 °C higher than 1850–1900, while global surface temperature in 2011–2020 was 1.09 °C higher than 1850–1900, with larger increases over land (1.59 °C) than over the ocean (0.88 °C). Many other variations have been observed in the climate system: increase of globally precipitation over land, changes in near-surface ocean salinity, global retreat of glaciers, warming and acidification of the oceans, increase of global mean sea level, changes in the land biosphere. In addition, climate change strongly affects climate extreme events, like heatwaves, heavy precipitation, droughts, and tropical cyclones.

Assuming that climate change is undoubtedly caused by anthropogenic activity, an overview of the main human activities that can affect the Climate System is provided below, which assume great potential in terms of climate change mitigation actions.

Typically, burning of fossil fuels and changes in land use are the major drivers that increase the GHGs concentration in the atmosphere. Referring to CO_2 concentration, without any doubt, the major cause of its increase is the burning of coal, oil and gas, which yields both carbon dioxide and nitrous oxide. In addition, CO_2 derives from transportation, building heating and cooling and the manufacture of products (mainly cement). Deforestation has also a negative impact on CO_2 levels, since trees help to guarantee the Earth's energy balance by absorbing CO_2 from the atmosphere. Consequently, when they are cut down, the stored carbon is released, increasing the greenhouse effect. In addition, changes in land use (like deforestation, but also construction of roads, cities etc.) result in a change in the reflectivity of the Earth's surface, affecting local warming. Referring to nitrous oxide, as by burning fossil fuels, it is also emitted by agricultural activities, due to the widespread use of fertilizers. Otherwise, the livestock farming also contributes to methane growth, since animals like cows and sheep produce a significant amount of CH_4 when they digest food. Finally, fluorinated gases are emitted from equipment such as refrigerators, air conditioners, sprays, switchgears in electrical distribution systems, leading to a very strong warming effect. Clearly, these are only a limited part of the human activities that contribute to the increase of emissions. Typically, these activities are categorized according to the economic sector they belong to, allowing to understand the most impactful sectors.

The IPCC Sixth Assessment Report recognised five mains economic sectors to be responsible for anthropogenic GHGs emissions in atmosphere. In 2019, looking at direct emissions, 34% of them are related to the energy sector, followed by industry (24%), Agriculture, Forestry and Other Land Uses (22%), transportation (15%) and buildings (6%). It is worth noting that the energy sector includes the most impactful sub-sector, which is electricity and heat generation. This subsector can be reallocated to the previous macro-sectors that use the final energy, accounting for their indirect emissions. Thus, considering both direct and indirect emissions, industry rises to the first rank achieving 34%, followed by buildings reaching 16%, while the other categories maintain the previous percentages.

2.1.4 Impacts, Vulnerability, and Adaptability

Climate change impacts, adaptation and vulnerability are the focus of the Report of the Second IPCC Working Group (WGII), whose main findings are briefly summarized in the present subsection [4].

The impacts of climate change are growing more pervasive, affecting every region of the planet, thus making not only nature, but also human lives and infrastructure vulnerable. Although the adverse consequences of climate change have been clear since the First IPCC Assessment Report [6], their extent and magnitude appeared beyond expectation in the latest report, making them harder to manage.

In detail, a wide range of effects can be related to climate change, which both directly cause such impacts or facilitate their occurrence, affecting ecosystems on a global scale as well as the human system. However, it has been revealed that these effects do not homogeneously affect different regions of our planet. Accordingly, the Sixth IPCC Assessment Report (AR6) identifies 127 key risks related to the adverse impacts of climate change, resulting from the interaction between three factors:

- *climate hazard(s)*, defined as "the potential occurrence of a natural or human-induced physical event or trend that may cause loss of life, injury, or other health impacts, as well as damage and loss to property, infrastructure, livelihoods, service provision, ecosystems and environmental resources".
- *exposure*, defined as "the presence of people; livelihoods; species or ecosystems; environmental functions, services and resources; infrastructure; or economic, social or cultural assets in places and settings that could be adversely affected".
- *vulnerability*, defined as "the propensity or predisposition to be adversely affected and encompasses a variety of concepts and elements, including sensitivity or susceptibility to harm and lack of capacity to cope and adapt".

In detail, climate change affects all the ecosystems: terrestrial, freshwater, and oceanic, which have been shown to have already been altered. Likewise, it also afflicts the human system, having effects on three main aspects, explained below.

- Water scarcity and food production, which include general water availability, agricultural production, livestock productivity and health, reduced fisheries yields and aquaculture production.
- Human health and wellbeing, involving infectious diseases, heat and malnutrition, mental health, and displacement.
- Cities, settlements, and infrastructure, which include inland flooding and associated damages, flood/storm induced damages in coastal areas, damages by key economic sectors.

Research highlights that the vulnerability of both ecosystems and people—which is interdependent—is not homogeneous across the planet but differs both between regions and within regions. The reason is to be sought in intersecting socioeconomic development, unsustainable ocean and land use, inequity, marginalization, historical and ongoing patterns of inequity, and governance.

Currently, our society is still pursuing unsustainable development patterns, which are increasing the exposure of ecosystems and people to climate risks, leading the IPCC to identify risks in the short (2021–2040) and medium term (2041–2100), closely related to the adaptation and mitigation efforts adopted. Indeed, adaptation is a crucial point in both reducing exposure and vulnerability, which—along with hazards—determine risk. Clearly, both natural and human systems exhibit the capacity to adapt. In natural systems, adaptation includes autonomous adjustments (through ecological and evolutionary processes) to the current climate and its effects and can occur within certain limits. In human systems, adaptation constitutes the process of adapting to the current or projected climate and its effects to moderate damage or take advantage of beneficial opportunities. In detail, for each key risk, the IPCC report provides several possible adaptation strategies as well as their feasibility, emphasizing that their implementation depends on the capacity and effectiveness of governance.

To summarize the process explained in this section, human activities result in climate change, which determines impacts and risks (based on hazard, vulnerability, and exposure), which may exceed the adaptive capacity of natural and ecological systems, thus leading to damage and losses. Different strategies are available to increase the systems adaptability, whose implementation depends on the government policies but is crucial to support a climate resilient development.

2.1.5 Climate Change Mitigation

To limit the adverse effects of climate change, increasing in adaptation capacity of human and natural system is not sufficient, but mitigation strategies are required. In detail, mitigation is defined as "a human intervention to reduce emissions or enhance the sinks of greenhouse gases". Overall, mitigation strategies involve strategies to reduce GHGs emissions from energy production/use and land use, as well as strategies to remove emissions already emitted in the atmosphere through land use or other mechanisms, included artificial ones.

The IPCC Working Group III provides an in-depth assessment of different mitigation strategies, highlighting both their estimated potentials and costs [3]. Clearly, such strategies can be applied to all sectors responsible for GHGs emissions: energy; agriculture, forestry, and other land uses; buildings; transport; industry; and others.

2.2 The Role of the Built Environment and Future Challenges

The present chapter aims to introduce and discuss the complex role of the built environment in the climate change framework, with a special focus on buildings. First, the potential effects of climate change at the global scale and at the local scale on the urban environment are presented, and related problems are prospected. Then, an in-depth analysis is presented highlighting the expected effects at the building scale, as well as potential adaptation and mitigation strategies.

2.2.1 The Built Environment in a Changing Climate

Human-made structures and networks that surround them compose the built environment, which involves three main components, explained below:

- Buildings
- Infrastructure (roads, railways, bridges, and essential services including energy, communications, and water supply)
- Human-made landscape (parks and gardens, designed green infrastructure etc.).

The relationships between these components are complex and interrelated, as well as associated with social, environmental, and economical aspects.

Currently, the built environment is facing several troubles. Besides problems such as overpopulation, poverty, excessive energy consumption, and pollution, the daunting challenges related to global and local climate change are increasingly arising. In fact, the vulnerability of the built environment to climate change is now well documented, which either directly produces adverse effects or exacerbates effects that would occur anyway because of the aforementioned problems [7]. In addition, due to the complex relationships that govern the built environment, impacts extend far beyond the area directly affected by the climate risk [3]. The main hazards of climate change on the built environment are found to be:

- extreme temperatures;
- flooding and sea level rise;
- water scarcity and security.

Climate change results in a growth of global average temperatures, as well as an increase in frequency and magnitude of regional heat wave events, which strongly impact urban systems, especially in larger cities. In fact, cities already suffer from higher temperatures than suburbs and rural areas due to the urban heat island phenomenon. Such phenomenon will be further amplified by increasing heat waves, thus increasing urban overheating, with important consequences on energy demand and supply, environmental quality, heat-related mortality, and survivability [8].

A further climate hazard which could affect the built environment is urban flooding. Urban flooding risk is already strongly affected by urban expansion, which reduces the permeable surface area of the soil. In addition, changes in the frequency and intensity of rainfall, as well as rising sea levels due to climate change, further increase the flooding risk [9]. Obviously, cities located on low elevation coastal zone are more exposed to such risk [10].

Urban water scarcity is also influenced by climate drivers as well as urbanization processes. In fact, land use changes, migration to cities (the increasing urbanization), together with rising temperatures and drought, result in an extension of the gap between water demand and supply [11].

Therefore, urgent adaptation measures are required to increase the built environment resilience to the irreversible impacts of climate change, even because cities are often exposed to multiple climate hazards [12]. At the same time, urban settlements strongly influence climate change as they are recognized to be large emitters of CO_2, accounting for 70% of global CO_2-eq emissions. Therefore, proper mitigation strategies need to be implemented alongside adaptation strategies.

2.2.2 *Buildings: Current Emissions*

As explained in the previous section, researchers recognize the building sector among the largest emitters of GHGs. In detail, the IPCC Working Group III (WGIII) revealed the observed data concerning emissions in 2019, showing that the construction sector was responsible of 21% of global GHGs emissions that year, corresponding to $12GtCO_2eq$ [3]. Such emissions result from several components:

- direct emissions produced on-site, related to building operations (i.e., heating, cooling, ventilation, or air conditioning);
- indirect emissions, associated with fuel combustion for the off-site generation of electricity and heat, then consumed in the buildings as end-use;
- embodied emissions associated with the production of building materials.

GHGs emissions generated by the construction sector are mainly related to CO_2 emissions, while methane and nitrous oxide emissions are negligible. Therefore, looking only at CO2 emission, the construction sector reaches 31% of global values [3]. The most updated data observed in 2022 also support this trend, confirming the construction sector accountable for about one-third of total energy system emissions (33%). In detail, 26% of emissions relies upon the building operations (8% representing direct emissions and 18% indirect emissions), while 7% of emission are embodied emissions related to buildings materials [13, 14].

2.2.3 Buildings: Mitigation Strategies

The IPCC WGIII definitively makes clear the responsibilities of the construction sector in CO_2 emissions. Consequently, it highlights its high potential in terms of climate change mitigation, and thus also high potential for achieving the UN Sustainable Development Goals and the Paris Agreement [3].

Since emissions result from different factors (direct, indirect, and embodied emissions), mitigation strategies can be manifold. For instance, the AR6 identifies five main mitigation opportunities, highlighting both potential and costs:

- *Avoid demand for energy services*
  This strategy requires a more effective reorganization of built spaces, the use of shared homes and facilities, a reorganization of human activities (through realizations of co-working, teleworking).
- *Efficient lighting, appliances, and equipment*
  This strategy assumes the replacement of lights, appliances, and office equipment, including ICT (Information and Communication Technologies), water heating, and cooking technologies with more efficient solutions, allowing emissions to be reduced by up to 13%, although the cost-benefit ratio is often limited.
- *New buildings with high energy performance*
  This strategy assumes the construction of buildings with high energy performance, characterized by appropriate design, shape, high envelope performance and advanced HVAC technologies. However, their cost-effectiveness is often debated. Such strategy has very high potential in developing countries, where most of built area has yet to be realized. The construction of high-performance buildings allows to reduce consumption related to the building operation stage, thus limiting CO_2 direct emissions. However, embodied emissions in the building lifetime could grow, due to additional building material.
- *Onsite renewable production and use*
  This strategy requires the integration of renewable energy technologies, backed-up with demand-side flexibility and digitalisation measures, allowing to further reduce the direct emissions.
- *Improvement of existing building stock*
  This strategy requires the improvement of both efficiency and utilisation of existing building stock. On one hand, energy retrofit could be deep retrofit, which involves both envelope retrofit and systems replacement. On the other hand, a less-deep retrofit can be adopted, followed by switching to low-carbon district heating or replacing current HVAC with heat pumps connected to on-site renewable sources supported by energy storage.

In addition, a further strategy can be identified, related to the decarbonizing of building materials production. Likewise, decarbonization of the electricity sector also has extremely beneficial impacts on the construction sector, as it reduces indirect emissions.

2.2.4 Buildings: Climate Change Impacts

The construction sector is not only responsible for climate change, but it also suffers its adverse effects, as it alters the environmental conditions in which the building operates. Indeed, buildings are found to be affected by global warming during their lifespan because of the change in outdoor conditions, regardless of the efforts to reduce anthropogenic GHGs emissions.

The potential impacts of climate change on the built environment can be summarised into four categories [15], explained below:

- impacts on building structures, mainly related to environmental catastrophes (i.e., floods, landslides, storms, sea level rise, excess snow load);
- impacts on building construction (i.e., decay of fastening and water supply systems);
- impacts on building material properties (diminished performance of frost-resistance, UV-resistance, or insulation due to material decay);
- impacts on building indoor conditions or building energy use (increase in indoor temperatures and relative humidity levels).

Climate change has been shown to have significant effects on buildings, which can simultaneously result in increased GHG emissions if mitigation measures are not taken, creating an adverse loop. However, the extent of the literature on climate change adaptation measures still seems to be limited compared to the magnitude of the topic, mainly focusing on impacts related to rising temperatures [16]. Consequently, impacts on building energy use and indoor conditions is undoubtedly the most investigated category in literature. Global warming will impact cooling and heating needs, leading to a reduction in heating consumptions and an increase in cooling consumptions, resulting in an overall growth in total energy consumption. Clearly, the impacts of climate change are not homogeneous across the planet, therefore the extent of these variations depends on the location investigated [17]. The increase of cooling needs not only affects energy consumption (and thus direct emission of GHGs), but also influence the peaks in energy demand. In fact, the increase in both frequency and magnitude of extreme phenomena such as heatwaves can increase the risk of grid failures and supply interruptions [18]. Climate change also has a negative impact on indoor comfort conditions, leading to an increased risk of overheating [19], especially in high-performance buildings [20]. Buildings located in cities particularly suffer from this risk due to urban overheating [8].

2.2.5 Buildings: Adaptation Measures

The vulnerabilities and potential adaptation strategies for the built environment are presented and widely discussed in the IPCC AR6 WGII Report, which dedicates a section to buildings as well [4].

Strategies for adapting buildings to climate change are manifold and are unavoidably intertwined with mitigation strategies. In addition, drawing a distinction between adaptation and mitigation strategies is often tricky, as some adaptation strategies have beneficial effects on mitigation and vice versa [21]. However, adaptation plans are found to be less common than mitigation plans [22]. Typically, climate adaptation can be achieved by increasing the adaptive capacity of buildings to climate variations, ensuring user comfort during climate extremes. For instance, due to climate change, maintaining adequate temperatures inside buildings will be increasingly harder, leading to discomfort conditions. Therefore, greater energy demand will be required to meet the thermal needs (mainly cooling needs), leading to an increase of energy consumption and thus CO_2 direct emissions [23].

Clearly, the adoption of different strategies is closely related to the building type, as well as the building location. Adaptation strategies generally include reducing heating and cooling demand, but also improving appliance efficiency and using renewable energy systems [24]. The literature reveals that—among the most widely adopted strategies—there are passive measures that improve the building envelope, enhancing its thermal performance. These options include changes in the thermal mass of opaque envelope [25], the implementation of green roofs or green facades [26], the increase of insulation levels [27, 28]. The latter reduces the heat exchange between outdoor and indoor environment, and thus contributes to reducing both heating and cooling needs, ensuring adequate indoor comfort conditions. However—especially in warmer climates—increasing the level of insulation could lead to negative impacts by increasing cooling needs [29], if not combined with further strategies [30]. In this regard, adding solar shading (i.e., overhangs, louvres) can greatly contribute to reduce cooling heat loads by shielding the building from solar radiation thus reducing internal heat loads [31]. Likewise, increasing natural ventilation—especially in the night-time—can significantly help to reduce internal loads [32]. However, the benefits of natural ventilation may become less effective in the long term due to the average increase in outdoor temperatures [33]. Although these constitute the adaptation strategies proven to be most effective, many other options are available.

Such strategies are known as *hard adaptation* measures since they alter the structure of the buildings. Along with them, there are the so-called *soft adaptation* measures, which instead include new users' behaviour (i.e., new window opening strategies), as well as new patterns of building use (i.e., different working hours, different schools' hours, different thermostat setpoint for building systems). Clearly, these strategies cannot replace hard measures, and their robustness in terms of probability of implementation is quite challenging to assess [34, 35].

2.2.6 Building Energy Efficiency Policy Package

Since the publication of the IPCC Fifth Assessment Report, the complex relationship between climate change and built environment has been recognized. Consequently, there has been an extensive expansion of policies regarding this issue.

Two documents assumed a crucial role within this context: the 2030 Agenda for Sustainable Development (the SDGs) approved in September 2015 and the Paris Climate Agreement released in December 2015. Both agreements highlighted the key role of cities and communities in fighting climate change. With the Paris Agreement, member states agreed to limit the global temperature increase to 1.5 °C above pre-industrial levels, by proposing climate action plans. Cities and national authorities have been recognised as key actors in this process [36]. The 2030 Agenda for Sustainable Development is a United Nations effort that establishes a global plan of action to address the world's most pressing challenges (poverty, hunger, health, education), including 17 sustainable development goals (SDGs), each of which has specific targets and related measurable indicators. Among these goals, one is fully dedicated to "sustainable cities and communities" (SDG11), which is closely related to SDG 13, which concerns climate action. In fact, SDG11 aims to create sustainable cities and communities, and sustainable urbanization contributes to reducing the environmental impact of urban areas. In addition, SDG11 aims to achieve resilient communities, which are key to dealing with the impacts of climate change. Both goals emphasize the importance of international cooperation and partnerships to achieve their goals. Collaboration between local and national governments, the private sector, civil society, and communities is crucial to both addressing climate change and achieving sustainable cities.

At the building level, regulatory policy instruments assume a key role in the implementation of both mitigation and adaptation policies. In detail, the main policy packages include regulatory instruments, information instruments and incentive instruments, explained below [37]:

- *Building Energy Codes* are standards or regulations that set the energy performance of buildings standards. They establish minimum energy efficiency requirements that buildings must meet, involving both building envelope and systems. Building energy codes are found to be the most effective policy instruments for buildings.
- *Environmental certifications* are informational policy instruments. making clear their impact in terms of energy and performance. This contributes to keeping consumers more informed, providing incentives for more sustainable practices.
- *Tax incentives and financing* represent a further key policy element, as they stimulate adoption of regulations and help overcome market barriers (such as high investment costs and limited access to capital), as well as drive action beyond minimum standards. Such incentives can be not only financial, but also include other types, such as fee reductions and density bonuses.

In addition, further policies include the implementation of specific energy retrofit programs encouraging owners to improve the efficiency of their properties to increase the annual renewal rate [38], as well as encouraging the use of renewable energy, or imposing reporting requirements on building energy performance.

2.3 Future Weather Forecasts: Principles and Tools

A pivotal issue in the struggle against climate change concerns future climate predictions, essential to investigate impacts of global warming, as well as potential adaptation and mitigation strategies. The science involved in predicting future climate is the climate science, which has succeeded in simulating the Earth's climate system through climate modelling. Briefly, climate projections are generated by global climate models (GCMs), which are validated against past weather observations, and then forced by future emissions scenarios—associated with different potential world developments—to simulate future weather conditions. The GCMs output are available in terms of predicted variations in climate variables with respect to a reference period. However, such outputs show inadequate spatial and temporal resolution for direct application in building performance assessments, which require hourly weather data with a local spatial resolution. Over time, several methodologies have been developed to spatially and temporally downscale the GCMs outputs. In addition, some tools have been developed to easily create future climate files suitable to be directly used in the buildings assessments.

The following subsections provide an in-depth investigation of this process, briefly investigating: the theory behind climate modelling (Sect. 2.3.1), the theory behind climate modelling (Sect. 2.3.2), downscaling techniques (Sect. 2.3.3), and future weather generation files tools (Sect. 2.3.4).

2.3.1 Climate Modelling

Climate modelling is based on the adoption of numerical models, known as General Circulation Models or Global Climate Models, both abbreviated as GCMs. Such models simulate the Earth's climate through mathematical equations representing fundamental physical and chemical processes [39]. The processes refer to the major climate system components: the atmospheric component, oceanic component, land surface component, and the sea ice component.

The idea of simulating the climate system by solving numerical equations is not recent, but goes back to 1922, when Lewis Richardson published the book entitled "*Weather Prediction by Numerical Process*". Basically, his idea was to forecast the weather solving mathematical equations describing atmospheric flow by dividing the land into a grid of cells [40]. However, the lack of computational technologies did not allow him to pursue his idea. Since that time, several research teams have been established with the aim of studying climate modelling. The first GCM of the atmosphere (AGCM) was developed in 1956 by Norman Philipps [41], while Bryan and Cox created the first GCM of the oceans (OGCM) in 1967 [42]. The first general circulation climate model combining both atmospheric and oceanic processes has been developed by the National Oceanic and Atmospheric Administration (NOAA) Geophysical Fluid Dynamics Laboratory in Princeton in the 1960s, allowing researchers to find out how the ocean and atmosphere simultaneously interact to affect climate [43]. Still today, the models employed in the IPCC analyses are the so-called Coupled Atmosphere–Ocean General Circulation Models (AOGCM), which combine AGCMs, OGCMs, sea-ice models, and models of land-surface processes. Actually, IPCC assessments are not based on the outputs of a single GCM, but of an ensemble of GCMs involved in the so-called Coupled Model Intercomparison Project (CMIP), currently in its Sixth Update (CMIP6).

GCMs rely upon a basic concept, not far from Richardson's idea: the globe is divided into a three-dimensional cells grid, representing the Earth's surface, looking at geographic locations and at elevations. In each cell, mathematical equations—based on the fundamental laws of physics, fluid motion and chemistry—are solved for a given number of time-steps, and then the interactions with neighbouring cells are evaluated. Consequently, GCMs can be characterized by different spatial and temporal resolutions, affecting computational demands. The horizontal size of cells (measured in degrees of latitude/longitude or in km/miles) defines the spatial resolution of the model, while the size of the time-steps used in models defines the temporal resolution [44].

Over the decades, thanks to the development of computational technologies, the complexity of models has significantly increased, both in terms of spatial resolution and physical phenomena considered in simulations. Referring to the processes simulated into the GCMs, the IPCC Fifth Assessment Report (AR5) provides an overview of the evolution of the models' capabilities to combine different components of the climate system, starting from the 1970s, presented in Fig. 2.2. As shown, over time, not only the number of involved processes increased, but also the complexity in modelling the individual process (represented by the different size of the circles in the figure).

Meanwhile, GCMs have also experienced a significant increase in spatial resolution (both in horizontal resolutions and vertical levels). The First IPCC Assessment Report (FAR) was based on GCMs characterized by a spatial resolution of about 500 km, reduced to 250 km (Second IPCC Assessment Report, SAR), 180 km (Third IPCC Assessment Report, TAR), 110 km (Fourth IPCC Assessment Report, AR4). The vertical resolution has also decreased from 1 oceanic layer and 10–15 atmospheric layers (FAR) to 30 oceanic layers and 30 atmospheric layers (AR4). The AR5 and AR6 reports are based on further accurate GCMs, both in their complexity and in the representativeness of the key variables. In general, the validity of a model is assessed by simulating it in the past and comparing the outputs with observed historical climate data [45]. Although the reliability of GCMs has improved over time, researchers are aware that they are still characterized by significantly uncertainties. As such, to generate future projections, multi-model ensembles of GCMs are

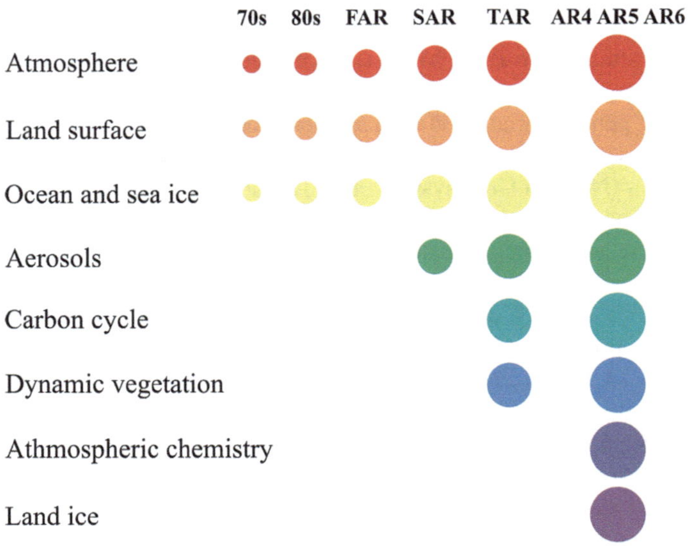

Fig. 2.2 Evolution in GCMs capabilities. The circles size represents the complexity in the process modelling

adopted, which allow to define both the uncertainty in projections and uncertainties in variations of initial conditions or parameterizations [46].

2.3.2 Climate Projections

GCMs represent the most effective tool available for simulating the climate on Earth. However, GCMs also allow researchers to assess the response of the climate system to the increase of greenhouse gases (GHG), thus making them the most effective tool to provide the so-called climate projections. The possible trend in GHGs levels constitutes different emission scenarios.

Climate projections represent a topic of great interest today, focus of several debates worldwide. Nevertheless, it is worth noting that the first crucial study on climate projections dates back to 1967, when Manabe and Wetherald evaluated the response of the atmosphere to the variation of CO_2 levels. Their aim was to assess changes in global average temperature when the radiative transfer of energy between surface and troposphere was modified by doubling the CO_2 levels. They outlined a temperature increase of 2.4 °C, setting the basis for all the future studies on global warming [47].

GCMs enable researchers to make climate projections, but it should not be neglected that they are characterized by a substantial level of uncertainty. Three main sources of uncertainty affected climate projections:

- emission scenarios uncertainty;
- models uncertainty;
- internal climate variability.

Consequently, they are likely to be associated with different confidence levels.

As discussed, the definition of emission scenarios is a pivotal element in climate projections, as GCMs are forced using such scenarios as inputs, with the aim to assess the response of the climate system. Over the years, the emission scenarios adopted by the IPCC have evolved from the IS92 scenarios to the current SSPs scenarios.

The first set of long-term emissions scenario was developed by the IPCC in 1990 (SA90 scenario) and adopted in the First IPCC report. Then, in 1992 a new set of scenarios (namely IS92), were presented in a supplementary report, involving six alternative scenarios, from IS92a to IS92f. Such scenarios involved an extensively range of assumptions on GHGs trends in the absence of climate policies, resulting in an order of magnitude range of potential GHG values [48]. Although the IPCC recommendation strongly suggested the use of all the six scenarios for climate projections to represent the emissions uncertainty, the IS92 has been the most adopted scenario in that period.

An assessment of IS92 scenarios was carried out in 1995, showing that substantial changes needed to be made. Consequently, in 1996 the IPCC Working Group III was tasked to create a new set of scenarios, released in the Special Report of Emission Scenarios in 2000 [49], and then adopted in the Third Assessment Report (TAR) in 2001. Such scenarios were further refined in the IPCC Fourth Assessment Report (AR4) published in 2007. Four narrative storylines were defined based on the relationship between the emission driving forces—not only greenhouse gases but also aerosol emissions—and their evolution over the twenty-first century. Climate policy-induced mitigation strategies were not considered. Consequently, each storyline was characterized by different demographic, social, economic, technological, and environmental developments. The four qualitative storylines generate four set of scenarios, namely the scenario families (A1, A2, B1, B2). Obviously, scenarios belonging to the same family share the same demographic, politico-societal, economic and technological storyline. The four storylines can be summarised as follows:

- A1 storyline: a future world of very rapid economic growth, global population that peaks in mid-century and declines thereafter, and rapid introduction of new and more efficient technologies;
- A2 storyline: a very heterogeneous world with continuously increasing global population and regionally oriented economic growth that is more fragmented and slower than in other storylines;
- B1 storyline: a convergent world with the same global population as in the A1 storyline but with rapid changes in economic structures toward a service and information economy, with reductions in material intensity, and the introduction of clean and resource efficient technologies;
- B2 storyline: a world in which the emphasis is on local solutions to economic, social, and environmental sustainability, with continuously increasing population (lower than A2) and intermediate economic development.

Six groups of scenarios were obtained from these four families. In detail, the A1 family involves three groups of scenarios, distinguished by different alternatives in the development of energy technologies: A1F1 (the fossil fuel intensive scenario), A1T (predominantly non-fossil fuel scenario) and A1B (balanced scenario). Otherwise, families A2, B1 and B2 include a single group of scenarios each. Finally, within each group and family, scenarios are distinguished in: (i) Harmonized Scenarios (HS), which share harmonized assumptions on global population, gross world product, and final energy; (ii) Overshoot Scenarios (OS), which consider uncertainties in the driving forces.

A further set of scenarios has been adopted in 2014 with the publication of the IPCC Fifth Assessment Report: the Representative Concentration Pathways (RCPs) [50]. Unlike the SRES scenarios which started from the socioeconomic context to project emission trends, the RCPs set the emission trends and the resulting radiative forcing a priori. Consequently, the same emission trend may result from different socioeconomic scenarios, and thus the RCP provides only one of the possible assumptions that lead to a specific radiative forcing. This is the reason behind the name "representative". Such set of scenarios involved four concentration pathways, labelled considering four radiative forcing values up to 2100 [51]. The radiative forcing is defined as "*the change in the net, downward minus upward, radiative flux at the tropopause or top of atmosphere due to a change in an external driver of climate change, such as, for example, a change in the concentration of carbon dioxide* (CO_2) *or the output of the Sun*" [50]. The four radiative forcings are represented by the number given in the name of each RCP, explained below.

- The RCP2.6 is the "peak" scenarios since the radiative forcing peaks at around 3 W/m^2 by the mid-century and then declines to 2.6 W/m^2 in 2100.
- The RCP4.5 is a stabilization scenario since the radiative forcing stabilizes at 4.5 W/m^2 in 2100.
- The RCP6.0 is a stabilization scenario since the radiative forcing stabilizes at 6.0 W/m^2 in 2100.
- The RCP8.5 is the worst scenario since the radiative force reaches 8.5 W/m^2 in 2100 and continues to grow, stabilizing only in 2250.

The latest set of scenarios was released in 2022 and then adopted in the last IPCC Sixth Assessment Report (AR6) [2]. The scenarios were called Shared Socio-Economic Pathways (SSPs) and included a wide range of scenarios. The SSPs are based on five narratives referred to socioeconomic assumptions, which involve socioeconomic challenges for both mitigation and adaptation [52]. The five storylines are briefly described below [53]:

- SSP1 represents a sustainable development scenario, characterised by low challenges to mitigation and adaptation;
- SSP2 represents a middle-of-the-road scenario, characterised by medium challenges to mitigation and adaptation;
- SSP3 represents the regional rivalry scenario, characterised by high challenges to mitigation and adaptation;

- SSP4 represents the inequality scenario, characterised by low challenges to mitigation and high challenges to adaptation;
- SSP5 represents the fossil fuels development scenario, characterised by high challenges to mitigation and low challenges to adaptation.

Such scenarios do not consider new climate policies further than those in place today, therefore they need to be combined with emission mitigation targets. Consequently, SSPs are coupled with the RCP projections, resulting in a SSPx-y scenario, where x refers to the SSP and y refers to the RCP (for instance SSP2-4.5). Therefore, each SSP scenario evaluates how the radiative forcings at 2100 set by the RCPs can be achieved based on their socioeconomic assumptions. The AR6 report climate projections are generated based on 5 scenarios: SSP1-1.9, SSP1-2.6, SSP2-4.5, SSP3-7.0, and SSP5-8.5.

2.3.3 Downscaling Techniques

As detailed in the previous paragraphs, the Global Climate Models—validated on the basis of past climate observations—can be forced with emission scenarios to obtain future climate projections [45]. However, the direct application of the outputs of GCMs in the building research field is not feasible, as they show both spatial and temporal inadequate resolution, along with bias in the data. Indeed, a GCM is typically characterized by a spatial resolution of about 100–300 km^2, with a monthly temporal resolution. By contrast, Building Performance Simulation (BPS) software require as boundary conditions hourly weather files (namely epw files), thus not consistent with the outputs of GCMs. Consequently, such outputs need to be spatially and temporally downscaled to be suitable for application in BPS tools.

It is worth noting that scaling the results of a GCM to a finer spatial and temporal resolution presumes the assumption that the local climate is a combination of large-scale climate features and local conditions (topography, water bodies, land surface properties), which are beyond the modelling capabilities of GCMs [54].

Over time, several methodologies have been developed to downscaled GCMs, which can be classified into three major approaches [55]:

- statistical downscaling;
- dynamical downscaling;
- hybrid downscaling.

It should be noted that any downscaling process involves several steps, each one relying on approximations and assumptions, thus increasing the level of uncertainty of climate projections [54].

Statistical downscaling

Statistical downscaling approaches are so labelled as they rely on the definition of empirical relationships between the large-scale climate features of GCMs and local climate variables. Such relationships are validated, and then adopted to predict future local climate variables based on the future variables predicted by the GCMs [54, 55]. Among the statistical techniques, two approaches appear to be the most adopted in literature, namely the imposed offset method and the stochastic weather model.

The *imposed offset method* combines future climate predictions from GCMs with typical current weather data. In detail, such method imposes the predicted climate information upon the current observed weather variables. This method has become most well-known in the form of the *morphing* method, developed by Belcher et al. [56]. Currently, it still appears one of the most widely adopted methods in the generation of weather files for BPS applications [17]. The morphing method is based on the transformation of hourly values of current climate variables to match the future changes of that variables predicted by the GCMs (or the Regional Climate Models RCMs). The current climate variables are historical observed variables that represent the present-day climate. Of course, the period of record of the current climate variables must be matched with the baseline considered to generate the predicted variations in the climate models, in order to provide reliable data [57].

Three algorithms are adopted to modify the current weather data, depending on the considered variable, namely Shift, Stretch and Combination of Shift and Stretch [56].

The Shift algorithm is an additive formulation. It adds an absolute monthly mean change (Δ_{xm}) predicted by a GCM or RCM to the hourly values of a weather variable in the weather file (x_0) for the month m, as described in Eq. (2.1).

$$x|_m = x_0 + \Delta x_m \tag{2.1}$$

The Stretch algorithm is a multiplicative formulation. It scales the hourly values of a variable in the weather file (x_0) by a predicted relative monthly mean change (α_m) for the month m, as described in Eq. (2.2).

$$x|_m = a_m \times x_0 \tag{2.2}$$

The Combination of shifting and stretching is a linear combination of the two previous algorithms. The hourly values of a weather variable in the weather file are both shifted by adding the predicted absolute monthly mean change (Δxm) and stretched by a predicted relative monthly mean change (α_m) for the month m, as described in Eq. (2.3).

$$x|_m = x_0 + \Delta x_m + a_m\left(x_0 - x_{0,m}\right) \tag{2.3}$$

where $x_{0,m}$ is the variable x_0 averaged over month m for all the considered averaging years of future data provided by climate models.

The main advantages and disadvantages of the imposed offset method are summarised in Table 2.1.

A further statistical downscaling approach is the stochastic weather model, developed by Van Paassen et al. [58] and Adelard et al. [59]. This technique allows to

Table 2.1 Main advantages and disadvantages of the imposed offset method, from Ref. [17]

Advantages	Disadvantages
• Minimal computing resources required [54] • Meteorologically consistent weather file generated, as the method is based on historical observations of weather [55, 57] • Typical reference weather year for the current climate normally available for building simulation [55]	• Applicability of future weather files constrained by baseline data availability [57], and high quality observed data may be unavailable [54] • Future increases in daily maximum and minimum temperature assumed to be the same as the predicted increases in average temperatures [55] • Possible change in solar insolation and wind speed ignored [55] • If the baseline data to be morphed is already in the form of a TRY (Test Reference Year), TMY (Typical Meteorological Year) there is also the inherent assumption that any climatic change that occurred between the baseline period of the weather file and the base- line period of the climate projections is negligible [57] • Currently observed relationships assumed to carry into the future [54]

generate future weather data based on an artificial meteorological database. In detail, statistical analyses of observed climate data are performed, adapting few independent weather variables to derive all the others. Thanks to this method, long synthetic time series can be generated, also filling the missing data. Table 2.2 summarises the main advantages and disadvantages of the stochastic weather model.

Dynamical downscaling

The dynamical approach allows to generate local climate data using a Regional Climate Model (RCM). Like GCMs, the RCMs are numerical model, but they are characterized by a finer resolution. Indeed, RCMs allow to simulate the Earth's climate adopting as boundary conditions the large-scale climate information derived from GCMs, combined with a more complex topography, the land-sea contrast,

Table 2.2 Main advantages and disadvantages of the stochastic weather model, from Ref. [17]

Advantages	Disadvantages
• Generation of future weather data also when historical data are not available [59] • Relationships between different weather variables accounted for in the same way as for baseline data [60] • Simulation of extreme weather conditions allowed [57] • Simulation of a wide range of feasible climate conditions feasible [57]	• Complex [55] • Computationally intensive [55] • Difficulties in accurately modelling many climatic variables [55]

Table 2.3 Main advantages and disadvantages of the dynamical downscaling, from Ref. [17]

Advantages	Disadvantages
• Based on consistent, physical mechanism [54] • Not constrained by historical record, thus novel scenarios can be simulated [54]	• Computationally intensive [54] • Large volumes of data required [54] • High level of expertise required [54] • Due to computational demands, RCMs are typically driven by only one or two GMC/emission scenario simulations [54] • May require further downscaling and bias correction of RCM outputs [54] • Affected by bias of forcing GCM [54]

surface heterogeneities, and detailed descriptions of physical processes. Consequently, a resolution of 20–50 km^2 can be achieved [54]. Therefore, in contrast to GCMs, which present such resolution that many processes must be approximated, RCMs allow for better modelling of the local landscape and local atmospheric processes. Obviously, since the RCM is nested into a GCM, the accuracy of an RCM is strictly related to the accuracy of the GCM adopted to force it.

An effort in assessing these uncertainties was made by combining and simulating different GCM-RCMs pairings, as was performed in the well-known ENSEMBLE [61] and EURO-CORDEX projects [62].

Like the other methods, dynamical downscaling presents both advantages and disadvantages, as shown in Table 2.3.

Hybrid downscaling

In some cases, when there is a need to reduce computational resources and the space to store data, it is possible to use hybrid downscaling. After obtaining the outputs from the various RCMs, these are scaled through statistical approaches.

2.3.4 Future Weather Generation File Tools

With the aim of generating future weather files readable to be used in BPS software, researchers could self-apply downscaling techniques to scalarize the future weather prediction provided by GCMs (or RCMs). However, such activity appears to be challenging and time-consuming, being more related to climatic science than to construction research field. Consequently, several tools have been developed over time to allow researchers to easily generate future weather files, which can be directly used in BPS tools. Although these tools certainly lead to a benefit by facilitating studies concerning the built environment in a changing climate, their adoption also brings disadvantages. Indeed, their use implies that the methodologies adopted in the studies—in terms of climate scenarios, time horizons of analysis, and downscaling techniques—are strongly correlated with the tools' capabilities. A striking example is the wide occurrence in the literature of manuscripts adopting the Special Report on Emissions Scenarios (SRES A2 as emission scenario to generate future weather files, since this represents the scenario adopted in the most common tool [17].

The next sections will provide a brief overview of the most commonly adopted tools, highlighting their advantages and disadvantages. Interestingly, the available tools are all based on statistical downscaling techniques. In detail, four of the analysed tools are based on the use of the morphing method, which is the easiest downscaling technique to be implemented. They are the CCWorldWeatherGen [63], the WeatherShift [64], and the Future Weather Generator [65]. Furthermore, one tool—Meteonorm [66]—is based on the stochastic weather method.

CCWorldWeatherGen

The CCWorldWeatherGenerator tool is a Microsoft Excel based tool, available free of charge. It has been developed in 2013 by the Sustainable Energy Research Group of Southampton University [63], although no longer maintained since 2017.

Such tool allows to create future weather file in an Energy Plus Weather (EPW) format, adopting the morphing method [56] as downscaling procedure. In detail, the climate predictions are generated with the General Circulation Model called Hadley Center Couple Model 3 (HadCM3) [67]—developed by the UK Office Met—forced with the emission scenario SRES A2 [49]. This model is selected among 29 other climate models, since it provides all the climate variables required by the morphing procedure [68]. The HadCM3 outputs are available on a monthly basis as relative changes concerning the 1961–1990 reference period, becoming the inputs to carry out the morphing procedure. Indeed, such inputs are superimposed on the baseline climate variables stored in a EPW weather file (baseline EPW). Obviously, to avoid inaccuracies, the reference EPW file should be based on the same 1961–1990 reference period. This procedure allows to create three future weather files:

- "2020s" weather file, representing the period 2001–2040;
- "2050s" weather file, representing the period 2041–2070;
- "2080s" weather file, representing the period 2071–2100.

WeatherShift

The WeatherShift TM is a web tool, available for a fee. It has been developed by Arup and Argos Analytics in 2016 [64]. Similar to the CCWorldWeatherGenerator tool, it allows to generate future weather file in an EPW format, adopting the morphing method as downscaling approach. However, unlike the CCWorldWeatherGen, WeatherShift is based on climate projections generated not only by a single GCM output, but by an ensemble of 14 GCMs. To provide future climate projections, these GCMs are forced with the RCP4.5 or the RCP8.5 emission scenarios adopted in the IPCC Fifth Assessment Report [50]. The GCMs outputs are provided as monthly variations adopting the 1976–2005 as reference period [69]. The tool provides future weather file for time periods of 20 years, from 2011 to 2100. Although, they are not available for all locations, but only for 259 cities. Using climate projections based on an ensemble of Global Circulation Models allows for the uncertainties associated with climate projections to be taken into account. As a result, the future weather files are associated with different levels of probability, corresponding to different warming percentile (10, 25, 50, 75, 90 and 95%).

Interestingly, the tool seems to have been updated with the new Shared Socioeconomic Pathways (SSPs), adopted in the IPCC Sixth Assessment Report. Nevertheless, no new documentation is provided on this update so far.

Future weather generator

The Future Weather Generator is Java app, available free of charge, developed in 2023 [65]. Like the previous tools, it allows to generate future weather file in an EPW format, adopting the morphing method as downscaling approach. The climate projections can be derived from different GCMs belonging to the CMIP6 models, allowing an ensemble of user selected models to be used. The baseline for the predicted monthly variations is the period 1985–2014. Such models are forced with the new Shared Socioeconomic Pathways (SSPs) adopted in the IPCC Sixth Assessment Report: SSP1-2.6, SSP2-4.5, SSP3-7.0, SSP5-8.5. The future weather files belong to two future timeframes:

- "2050s" weather file, representing the period 2036–2065;
- "2080s" weather file, representing the period 2066–2095.

Meteonorm

Meteonorm is a software that allows hourly weather data to be generated for any location worldwide, for historical, contemporary, or future scenarios [66].

Unlike the previous tool, this software adopts the stochastic weather method as downscaling approach. The climate projections are based on a selection of ten CMIP5 models, forced with the Representative Concentration Scenarios RCP2.6, RCP4.5 and RCP8.5, adopted in the IPCC Fourth Assessment Report. The models are average for three future time periods: 2011–2030, 2046–2065, 2080–2099. Then, the results are linearly interpolated in between, to generate future weather files, with 10-years intervals (from 2020 to 2100).

References

1. United Nations, *United Nations Framework Convention on Climate Change*
2. Intergovernmental Panel on Climate Change (IPCC), *Climate Change 2021: The Physical Science Basis. Contribution of Working Group I to the Sixth Assessment Report of the Intergovernmental Panel on Climate Change* (Cambridge University Press, Cambridge, United Kingdom and New York, NY, USA, 2023). ISBN 9781009157896
3. Intergovernmental Panel on Climate Change (IPCC), *Climate Change 2022: Mitigation of Climate Change. Contribution of Working Group III to the Sixth Assessment Report of the Intergovernmental Panel on Climate Change.* (Cambridge University Press, Cambridge, UK and New York, NY, USA, 2022)
4. Intergovernmental Panel on Climate Change (IPCC), *Climate Change 2022—Impacts, Adaptation and Vulnerability* (Cambridge University Press, 2023). ISBN 9781009325844

5. J.T. Kiehl, K.E. Trenberth, Earth's annual global mean energy budget. Bull. Am. Meteorol. Soc. **78**, 197–208 (1997). https://doi.org/10.1175/1520-0477(1997)078%3c0197:EAGMEB%3e2.0.CO;2
6. Intergovernmental Panel on Climate Change (IPCC), *Climate Change: The IPCC Scientific Assessment* (Cambridge University Press, 1990)
7. B. Güneralp, İ Güneralp, Y. Liu, Changing global patterns of urban exposure to flood and drought hazards. Glob. Environ. Chang. **31**, 217–225 (2015). https://doi.org/10.1016/j.gloenvcha.2015.01.002
8. J. Feng, K. Gao, H. Khan, G. Ulpiani, K. Vasilakopoulou, G.Y. Yun, M. Santamouris, Overheating of cities—magnitude, characteristics, impact, mitigation and adaptation and future challenges. Annu. Rev. Environ. Resour.**48** (2023)
9. L. Alfieri, S. Cohen, J. Galantowicz, G.J.-P. Schumann, M.A. Trigg, E. Zsoter, C. Prudhomme, A. Kruczkiewicz, E. Coughlan de Perez, Z. Flamig, et al., A global network for operational flood risk reduction. Environ. Sci. Policy **84**, 149–158 (2018). https://doi.org/10.1016/j.envsci.2018.03.014
10. G.L. Cozannet, R.J. Nicholls, G. Durand, A.B.A. Slangen, D. Lincke, A. Chapuis, Adaptation to multi-meter sea-level rise should start now. Environ. Res. Lett. **18**, 091001 (2023). https://doi.org/10.1088/1748-9326/acef3f
11. S.D. Crausbay, J. Betancourt, J. Bradford, J. Cartwright, W.C. Dennison, J. Dunham, C.A.F. Enquist, A.G. Frazier, K.R. Hall, J.S. Littell et al., Unfamiliar territory: emerging themes for ecological drought research and management. One Earth **3**, 337–353 (2020). https://doi.org/10.1016/j.oneear.2020.08.019
12. S.B. Guerreiro, R.J. Dawson, C. Kilsby, E. Lewis, A. Ford, Future heat-waves, droughts and floods in 571 European cities. Environ. Res. Lett. **13**, 034009 (2018). https://doi.org/10.1088/1748-9326/aaaad3
13. IEA; IRENA; UN Climate Change High Level Champions, *Breakthrough Agenda Report 2023*, Paris (2023)
14. IEA. *Tracking Clean Energy Progress 2023*, Paris (2023)
15. I. Andrić, M. Koc, S.G. Al-Ghamdi, A review of climate change implications for built environment: impacts, mitigation measures and associated challenges in developed and developing countries. J. Clean. Prod. **211**, 83–102 (2019). https://doi.org/10.1016/j.jclepro.2018.11.128
16. A.E. Stagrum, E. Andenæs, T. Kvande, J. Lohne, Climate change adaptation measures for buildings-a scoping review. Sustainability (Switzerland) **12** (2020). https://doi.org/10.3390/su12051721
17. L.M. Campagna, F. Fiorito, On the impact of climate change on building energy consumptions: a meta-analysis. Energies (Basel) **15**, 354 (2022). https://doi.org/10.3390/en15010354
18. M. Auffhammer, P. Baylis, C.H. Hausman, Climate change is projected to have severe impacts on the frequency and intensity of peak electricity demand across the United States. Proc. Natl. Acad. Sci. **114**, 1886–1891 (2017). https://doi.org/10.1073/pnas.1613193114
19. D. Fosas, D.A. Coley, S. Natarajan, M. Herrera, M. Fosas de Pando, A. Ramallo-Gonzalez, Mitigation versus adaptation: does insulating dwellings increase overheating risk? Build. Environ.**143**, 740–759 (2018). https://doi.org/10.1016/j.buildenv.2018.07.033
20. T. Picard, T. Hong, N. Luo, S.H. Lee, K. Sun, Robustness of energy performance of zero-net-energy (ZNE) homes. Energy Build. **224**, 110251 (2020). https://doi.org/10.1016/j.enbuild.2020.110251
21. H. Yassaghi, S. Hoque, An overview of climate change and building energy: performance, responses and uncertainties. Buildings (2019)
22. D. Reckien, M. Salvia, O. Heidrich, J.M. Church, F. Pietrapertosa, S. De Gregorio-Hurtado, V. D'Alonzo, A. Foley, S.G. Simoes, E. Krkoška Lorencová et al., How are cities planning to respond to climate change? Assessment of local climate plans from 885 cities in the EU-28. J. Clean. Prod. **191**, 207–219 (2018). https://doi.org/10.1016/j.jclepro.2018.03.220
23. D.H.W. Li, L. Yang, J.C. Lam, Impact of climate change on energy use in the built environment in different climate zones—a review. Energy **42**, 103–112 (2012). https://doi.org/10.1016/j.energy.2012.03.044

24. Z. Ren, Z. Chen, X. Wang, Climate change adaptation pathways for Australian residential buildings. Build. Environ. **46**, 2398–2412 (2011). https://doi.org/10.1016/j.buildenv.2011.05.022
25. A.U. Din, L. Brotas, The impacts of overheating mitigation within the life cycle carbon of dwellings under UK future climate. Proc. Environ. Sci. **38**, 836–843 (2017). https://doi.org/10.1016/j.proenv.2017.03.169
26. R. Castiglia Feitosa, S.J. Wilkinson, Small-scale experiments of seasonal heat stress attenuation through a combination of green roof and green walls. J. Clean Prod. **250**, 119443 (2020). https://doi.org/10.1016/j.jclepro.2019.119443
27. A. Invidiata, E. Ghisi, Impact of climate change on heating and cooling energy demand in houses in Brazil. Energy Build. **130**, 20–32 (2016). https://doi.org/10.1016/j.enbuild.2016.07.067
28. S.M. Iles, M. Seddiki, L. Kadi, Assessment of climate change adaptation measures in tertiary buildings under the Passivhaus Enerphit standard within the Mediterranean context. Int. J. Sustain. Build. Technol. Urban Dev. **14**, 169–190 (2023)
29. M.S. Fernandes, E. Rodrigues, A.R. Gaspar, J.J. Costa, Á. Gomes, The impact of thermal transmittance variation on building design in the Mediterranean region. Appl. Energy **239**, 581–597 (2019). https://doi.org/10.1016/j.apenergy.2019.01.239
30. S. Gilani, W. O'Brien, Natural ventilation usability under climate change in Canada and the United States. Build. Res. Inf. **49**, 367–386 (2021). https://doi.org/10.1080/09613218.2020.1760775
31. S.M. Al-Masrani, K.M. Al-Obaidi, N.A. Zalin, M.I. Aida Isma, Design optimisation of solar shading systems for tropical office buildings: challenges and future trends. Sol. Energy **170**, 849–872 (2018). https://doi.org/10.1016/j.solener.2018.04.047
32. B. Gao, X. Zhu, J. Ren, J. Ran, M.K. Kim, J. Liu, Multi-objective optimization of energy-saving measures and operation parameters for a newly retrofitted building in future climate conditions: a case study of an office building in Chengdu. Energy Rep. **9**, 2269–2285 (2023). https://doi.org/10.1016/j.egyr.2023.01.049
33. M.M. Osman, H. Sevinc, Adaptation of climate-responsive building design strategies and resilience to climate change in the hot/arid region of Khartoum, Sudan. Sustain. Cities Soc. **47**, 101429 (2019). https://doi.org/10.1016/j.scs.2019.101429
34. D. Coley, T. Kershaw, M. Eames, A Comparison of structural and behavioural adaptations to future proofing buildings against higher temperatures. Build. Environ. **55**, 159–166 (2012). https://doi.org/10.1016/j.buildenv.2011.12.011
35. P. Jafarpur, U. Berardi, Effects of climate changes on building energy demand and thermal comfort in Canadian office buildings adopting different temperature setpoints. J. Build. Eng. **42**, 102725 (2021). https://doi.org/10.1016/j.jobe.2021.102725
36. United Nations. *Paris Agreement to the United Nations Framework Convention on Climate Change*
37. International Energy Agency (IEA). *The Energy Efficiency Policy Package: Key Catalyst for Building Decarbonisation and Climate Action*, Paris (2023)
38. *A Renovation Wave for Europe—Greening Our Buildings, Creating Jobs, Improving Lives.* https://eur-lex.europa.eu/resource.html?uri=cellar:0638aa1d-0f02-11eb-bc07-01aa75ed71a1.0003.02/DOC_1&format=PDF. Accessed 3 March 2022
39. S. Trzaska, E. Schnarr, *A Review of Downscaling Methods for Climate Change Projections. African and Latin American Resilience to Climate Change (ARCC)*, United States (2014)
40. L. Richardson, *Weather Prediction by Numerical Process* (Cambridge University Press, Cambridge, 1922) (reprinted NY, Dover, 1965)
41. N.A. Phillips, The general circulation of the atmosphere: a numerical experiment. Q. J. R. Meteorol. Soc. **82**, 123–164 (1956). https://doi.org/10.1002/qj.49708235202
42. K. Bryan, M.D. Cox, A numerical investigation of the oceanic general circulation. Tellus **19**, 54–80 (1967). https://doi.org/10.1111/j.2153-3490.1967.tb01459.x
43. S. Manabe, K. Bryan, Climate calculations with a combined ocean-atmosphere model. J. Atmos. Sci. **26**, 786–789 (1969)

44. *GFDL Geophysical Fluid Dynamics Laboratory Climate Modeling*. https://www.gfdl.noaa.gov/climate-modeling/. Accessed 10 Nov 2023
45. S.M. Uppala, P.W. Kallberg, A.J. Simmons, U. Andrae, V.D. Bechtold, M. Fiorino, J.K. Gibson, J. Haseler, A. Hernandez, G.A. Kelly et al., The ERA-40 re-analysis. Q. J. R. Meteorol. Soc. **131**, 2961–3012 (2005)
46. R. Knutti, R. Furrer, C. Tebaldi, J. Cermak, G.A. Meehl, Challenges in combining projections from multiple climate models. J. Clim. **23**, 2739–2758 (2010). https://doi.org/10.1175/2009JCLI3361.1
47. S. Manabe, R.T. Wetherald, Thermal equilibrium of the atmosphere with a given distribution of relative humidity. J. Atmos. Sci. **24**, 241–259 (1967). https://doi.org/10.1175/1520-0469(1967)024%3c0241:TEOTAW%3e2.0.CO;2
48. J. Leggett, W.J. Pepper, R.J. Swart, J. Edmonds, L.G. Filho Meira, I. Mintzer, M.X. Wang, J. Watson, *"Emissions Scenarios for the IPCC: An Update", Climate Change 1992: The Supplementary Report to The IPCC Scientific Assessment* (1992)
49. N. Nakicenovic, J. Alcamo, A. Grubler, K. Riahi, R. Roehrl, H.-H. Rogner, et al., *Special Report on Emissions Scenarios (SRES), A Special Report of Working Group III of the Intergovernmental Panel on Climate Change* (2000)
50. IPCC. Near-term climate change: projections and predictability, in *Climate Change 2013: The Physical Science Basis. Contribution of Working Group I to the Fifth Assessment Report of the Intergovernmental Panel on Climate Change* (2013)
51. D.P. van Vuuren, J. Edmonds, M. Kainuma, K. Riahi, A. Thomson, K. Hibbard, G.C. Hurtt, T. Kram, V. Krey, J.-F. Lamarque et al., The representative concentration pathways: an overview. Clim. Change **109**, 5–31 (2011). https://doi.org/10.1007/s10584-011-0148-z
52. B.C. O'Neill, E. Kriegler, K. Riahi, K.L. Ebi, S. Hallegatte, T.R. Carter, R. Mathur, D.P. van Vuuren, A new scenario framework for climate change research: the concept of shared socioeconomic pathways. Clim. Change **122**, 387–400 (2014). https://doi.org/10.1007/s10584-013-0905-2
53. K. Riahi, D.P. van Vuuren, E. Kriegler, J. Edmonds, B.C. O'Neill, S. Fujimori, N. Bauer, K. Calvin, R. Dellink, O. Fricko et al., The shared socioeconomic pathways and their energy, land use, and greenhouse gas emissions implications: an overview. Glob. Environ. Chang. **42**, 153–168 (2017). https://doi.org/10.1016/j.gloenvcha.2016.05.009
54. S. Trzaska, E. Schnarr, *A Review of Downscaling Methods for Climate Change Projections* (2014)
55. L. Guan, Preparation of future weather data to study the impact of climate change on buildings. Build. Environ. **44**, 793–800 (2009). https://doi.org/10.1016/j.buildenv.2008.05.021
56. S.E. Belcher, J.N. Hacker, D.S. Powell, Constructing design weather data for future climates. Build. Serv. Eng. Res. Technol. **26**(1), 49–61 (2005). https://doi.org/10.1191/0143624405bt112oa
57. M. Herrera, S. Natarajan, D.A. Coley, T. Kershaw, A.P. Ramallo-González, M. Eames, D. Fosas, M. Wood, A review of current and future weather data for building simulation. Build. Serv. Eng. Res. Technol. **38**, 602–627 (2017). https://doi.org/10.1177/0143624417705937
58. A.H.C. van Paassen, Q.X. Luo, Weather data generator to study climate change on buildings. Build. Serv. Eng. Res. Technol. Int. J. **23**(4), 251–258 (2002)
59. L. Adelard, H. Boyer, F. Garde, J.C. Gatina, Detailed weather data generator for building simulations. Energy Build. **31**(1), 75–88 (2000)
60. U. Berardi, P. Jafarpur, Assessing the impact of climate change on building heating and cooling energy demand in Canada. Renew. Sustain. Energy Rev. **121** (2020). https://doi.org/10.1016/j.rser.2019.109681
61. P. van der Linden, J.F.B. Mitchell, *ENSEMBLES: Climate Change and Its Impacts: Summary of Research and Results from the ENSEMBLES Project* (2009)
62. D. Jacob, J. Petersen, B. Eggert, A. Alias, O.B. Christensen, L.M. Bouwer, A. Braun, A. Colette, M. Déqué, G. Georgievski et al., EURO-CORDEX: new high-resolution climate change projections for European impact research. Reg. Environ. Change **14**, 563–578 (2014). https://doi.org/10.1007/s10113-013-0499-2

63. M.F. Jentsch, P.A.B. James, L. Bourikas, A.B.S. Bahaj, Transforming existing weather data for worldwide locations to enable energy and building performance simulation under future climates. Renew. Energy **55**, 514–524 (2013). https://doi.org/10.1016/j.renene.2012.12.049
64. *Weather Shift*. https://weathershift.com/. Accessed 15 Nov 2023
65. E. Rodrigues, M.S. Fernandes, D. Carvalho, Future weather generator for building performance research: an open-source morphing tool and an application. Build. Environ. **233** (2023). https://doi.org/10.1016/j.buildenv.2023.110104
66. Meteotest Meteonorm Global Meteorological Database
67. Met Office HadCM3: Met Office Climate Prediction Model. https://www.metoffice.gov.uk/research/approach/modelling-systems/unified-model/climate-models/hadcm3. Accessed 15 Nov 2023
68. University of Southampton CCWeatherWorldGen Technical Reference Manual. Accessed 1 Nov 2021
69. R. Dickinson, B. Brannon, Generating the future weather files for resilience, in *International Conference on Passive and Low Energy Architecture* (2016)

Chapter 3
Responsive Envelope Technologies

Abstract This chapter aims to provide a thorough description of the responsive envelope topic from a theoretical point of view to allow a clear understanding of the concepts proposed in the following sections. Firstly, this chapter begins with a detailed overview of the main available classification systems and nomenclatures adopted in this research field. Subsequently, this chapter explores the different functioning criteria of the main responsive technologies currently available. Each functioning criterion is described in a dedicated subchapter where the most promising and representative responsive technologies are described—in a specific section—from a technological point of view.

Keywords Responsive · Adaptive · Envelope · Façade · Architectural engineering

The framework described in the introductory section explains the increasing awareness among policymakers regarding the need to improve the energy efficiency in buildings. Nowadays, researchers are supporting this new awareness—and the subsequent innovations and changes—focusing their attention on new building technologies. In particular, among the different innovative solutions proposed in the last years, this work is focused on responsive envelopes, namely those systems that aim to improve the energy efficiency and the users' need through reversible changes of their properties to adapt themselves to changing boundary conditions [1]. Depending on the specific technology, these systems can act on the control of the solar radiation—which affects both solar loads and daylighting—, on thermal storage and insulation, on ventilation, and on other physical phenomena to adapt the envelope to different external stimuli [2]. This chapter provides, firstly, an in-depth description of the available nomenclature and classification systems to allow a clear understanding of the proposed concepts; afterwards, the different functioning criteria are described before analysing in depth each specific technology.

F. Carlucci et al., *Responsive Envelopes and Climate Change*, Digital Innovations in Architecture, Engineering and Construction,
https://doi.org/10.1007/978-3-031-58101-4_3

3.1 Classification Systems

The fast spread of these technologies and concepts has led, in recent years, to a confusing use of many similar terms such as adaptive, responsive, kinetic, etc. Nevertheless, the first theorisations of these concepts find their roots at the end of the last century and can help to grasp the meaning of these terms. One of the first conceptualization of responsiveness was provided by Negroponte [3] who defined architecture—in particular housings—as "unresponsive to the needs and desires of their users". The author identified the causes in the design process and proposed, as viable solution, the participation of the users in the design and control of the building. This first architectural theorisation—apparently not involved in designing building envelope—was then declined in 1981, from a more technical point of view, in the paper "A wall for all seasons" [4]. In this article the British architect Mike Davies conceived a new envelope as a thin multifunctional transparent layer to satisfy different performance requirements removing "the distinction between solid and transparent". Moreover, according to the author, this wall would have acted as a filter that continuously adapt to the surroundings situations. Notwithstanding this theorisation appears close to the current connotation of the topic, the real spread and definition of these concepts started in the twenty-first century when the increasing development of new technologies and materials supported the implementation of these theories defining them more clearly. In this renovated context, responsive [5], adaptive [6], kinetic [7], dynamic [8], intelligent [9], smart [10], and switchable [11] are the most common adjectives associated to envelopes with time-varying properties. Despite all these terms seem interchangeable, each one focuses on specific features which should be properly considered; a specific study [12] theorised and visualised (Fig. 3.1) the relationships between the different adopted nomenclatures which are reported in the subsequent paragraphs.

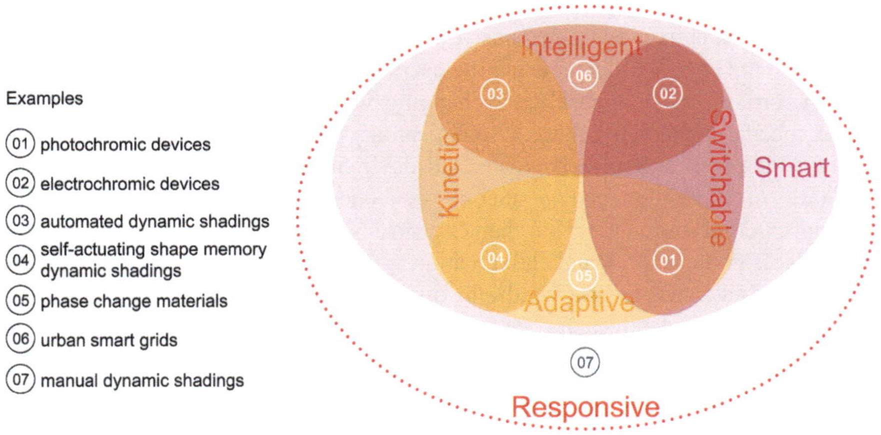

Fig. 3.1 Main adopted nomenclatures, possible overlapping, and examples [12]

In the author's opinion, the widest definition is given by the term "responsive" proposed in the Annex 44 of the IEA (International Energy Agency)—Energy Conservation in Buildings and Community Systems Programme (IEA-ECBCS) [13]. According to this source, the Responsive Building Elements (RBEs) were defined as components that can react in a predefined way in response to external stimuli or to users' interaction to control the transfer and storage of heat, light, air, and water [14].

A slightly different definition can be adopted for adaptiveness; Loonen [15] considers Climate Adaptive Building Shells (CABS) those envelopes that can change their properties repeatedly and reversibly according to a variation of the boundary conditions to improve the Indoor Environmental Quality (IEQ) and reduce the energy demand [15]. In the writer's opinion, this definition includes few technologies which are not literally "adaptive" such as electrochromic glazing; on the contrary, the adaptive systems should include only self-triggering devices with autonomous sensing and actuating systems. According to this view, intelligent and adaptive systems are different as better explained in the following sections. Nevertheless, currently both interpretations and meanings can be easily found in the scientific field. Several other definitions were theorised for the so-called AFs (adaptive facades) and the main emerging distinctive features were essentially the real time adaptation to boundary conditions.

While the previous definitions are usually considered extremely similar, the term kinetic is commonly characterised by a slightly different meaning. The Acclimated Kinetic Envelopes (AKEs) were defined by Wang et al. as envelopes that react to external stimuli with visible physical changes [16]; therefore, this definition is mainly focused on macroscopic—usually the movement of building components—rather than microscopic variations. The term dynamic is usually adopted as synonym of kinetic even though dynamic is sometime used also in different contexts—with larger patterns of meanings—such as dynamic glazing.

Instead, the main feature of intelligent devices is their organization in specific components for perception, cognition, and action to whom correspond sensors, controllers, and actuators [17]. Hence, thanks to this structure, these devices can change their properties through automated adjustments [18].

Another extremely diffuse term is "smart" which is usually associated to both materials and technologies; it is, apparently, very similar to "intelligent" but includes further shades of meaning. Smart materials can be classified in property exchanging or energy exchanging [19]; the former can change their thermal, mechanical, magnetic, optical, or electrical properties (e.g., photochromic films) while the latter can convert energy (e.g., photovoltaic). Although, the term smart can also be applied to larger contexts such as buildings, grids, or cities when it is referred mainly to advanced control systems for interconnected operability [20]. This brief description highlights a wide possible use of the term smart which can be used to describe both intelligent and adaptive systems [21].

Finally, the last term considered in this brief review is "switchable" which is generally adopted in the windows and glazing field. The switchable glazing can modulate or switch its properties according to external stimuli; depending on its functioning it can be classified as actively or passively controlled [11]. The former class includes all those systems that can be extrinsically controlled such as the electrochromic glazing; while passively controlled systems change their properties according to their intrinsic structure such as, for example, the photochromic glazing, as better explained in the dedicated section. This brief list of the nomenclature adopted in this research field hints the complexity of the topic and the broad number of different contributions and studies conducted in the last years. The spread of these studies and the subsequent large availability of disparate technologies were supported by the development of different classification systems based on different sorting criteria. For example, one of the most recent systems [22] is based basically on the opinions of the main stakeholders. Indeed, after interviewing architects, engineers, manufacturers, contractors, and operators, the authors defined four main families—dynamic shadings, chromogenic façade, solar active façade, and active ventilative façade—composed by 11 technologies overall. A further level of description is provided to each technology which could be classified based on its application-purpose (privacy, insulation, etc.), control type (manual, automated, etc.), building type (residential, etc.), and technology-material (wood, suspended particles, etc.).

A different approach was considered by Loonen et al. who provided two different classification systems. The first one [1] considers the physics of the system (thermal, optical, air flow, and electrical) and then identifies its responsiveness time scale (seconds, minutes, hours, seasons), its scale of adaptation (micro or macro), its control type (intrinsic or extrinsic), and its typology (built, subsystems, full scale, or reduced scale prototype). The second one [23] is a review of the first proposal and starts from the purpose-goal of the system (thermal comfort, indoor air quality, visual performance, acoustic quality, energy generation, personal control), then considers the responsive function (e.g., modulate solar gains), the operation (intrinsic or extrinsic), the technologies-materials adopted (e.g., Phase Change Material, switchable glazing), the response time, the spatial scale (building material, façade element, wall, fenestration, roof, whole building), the visibility (no, low, high) and, finally, the degree of adaptability (on/off or gradual). This last classification system, more thoroughly described in the next section, is considered extremely complete and, therefore, is adopted in this book as main reference.

Referring to AKEs, the classification proposed by Wang et al. [16] could be considered as a possible approach. Firstly, the authors identify the relation between climate source and façade (solar responsive, air flow responsive, others). The following classification level considers the parameters that control the devices: solar heat, daylight, solar electricity—for the solar responsive systems—, natural ventilation, or wind electricity—for the air flow responsive—, and precipitation or air temperature, for the "others" class.

While these systems can describe the largest number of responsive technologies, other classification systems can be adopted specifically for certain devices. For example, with regards to kinetic systems, another classification system [24] takes into account four main classification levels: kinematic (limited, medium, major, variable), control technique (direct-responsive, internal or direct, responsive indirect), system configuration (embedded, dynamic), control limit (minor, medium, significant, variable), and cost (small, medium, big, huge). Combining these four criteria, different categories—skin unit systems, retractable systems, revolving buildings, and biomechanical buildings—can be identified to describe the considered system.

Instead, a different approach was adopted by Ochoa and Capeluto [25] who consider three main categories (input elements, control processing elements, actuating elements) to classify the devices. Starting from these main classes, the authors reach more than 40 possible combinations branching each class in different subclasses (category, design variable, sub-variable, common values).

3.2 Functioning Criteria

The classification system proposed by Loonen et al. [23] can be adopted as starting point for understanding the functioning criteria of the responsive devices. Looking at the purpose-goal class, it can be noted that responsive systems can act on a multiphysics domain working—simultaneously or singularly depending on the technologies—on thermal comfort, indoor air quality, visual performance, acoustic comfort, and energy generation. Depending on the considered domain, the responsiveness can be expressed in several ways; for example, to improve the thermal comfort the responsive device can act on the solar radiation (allow/modulate/reject the solar gains), on the heat flux (increase/decrease the thermal transmittance), and/or on the thermal inertia (release/absorb the latent heat). The device can express its responsiveness in three main ways: intrinsically, extrinsically, or in a combined way. The intrinsic or passive control allows the system to spontaneously self-adjust its properties according to variations of the environmental parameters. On the contrary, extrinsic (active) control is based on a sensor-processor-actuator architecture that allows the responsiveness of the system. Finally, few technologies—such as, for example, the thermo-electrochromic windows [26]—merge intrinsically and extrinsically controlled technologies in a single device obtaining a combined control where benefits and drawbacks of each system are balanced. On the one hand, the intrinsically controlled systems are "naturally" responsive, hence no specific additional sensing-processing-acting systems are required; this feature allows to significantly reduce, or avoid, the maintenance and the risk of breakage of the device. On the other hand, no external control can be applied on these systems; therefore, users cannot interact or override the status of the device and it is fundamental to properly design the component for each specific application (accounting for climate, exposure, building use, etc.) to avoid loss in energy and comfort performances.

On the contrary, the extrinsically controlled systems can be operated in a more flexible manner and can be easily managed by complex control algorithms that can optimise the performance of the envelope component accounting for different aspects (e.g., daylight, solar gains, occupancy, users' need etc.). For example, an electrochromic window can be connected to a daylighting sensor to improve the users' comfort but can be manually overridden for privacy reasons or for specific uses of the room (screen presentations, exhibitions, etc.). Moreover, the same technology can be connected to a different sensor (e.g., temperature, solar radiation, etc.), can be managed through model predictive control technique, or can be optimised during its service life simply adjusting the control algorithm to completely change its behaviour. Nevertheless, these devices require higher investment and maintenance costs related to the sensing-processing-acting systems.

Regardless of the control of the device, the transition between different states can be described through the response time and the degree of responsiveness. The former represents the readiness of the system and usually ranges from seconds (real time response) to hours depending on the implemented technology and on the designer's approach. While the response time provides a description of the sensitivity referred to the time scale, the degree of responsiveness expresses the sensitivity with regards to the variation of the boundary conditions. If the device can switch between two states (on–off) the degree of responsiveness is very low as it cannot smoothly accommodate the changing boundary conditions; on the contrary, modulating the properties between different states (gradual modulation) allows to respond to the variations more accurately as shown in Fig. 3.2.

The transition between different states can take place at a molecular scale or at a macroscopic scale depending on the technology adopted. The former regards the material itself and usually affects the thermophysical or optical properties of the device (e.g., Phase Change Materials, photochromic glazing, etc.) whilst the macroscopic variations regard the whole system (e.g., dynamic shading system). While the macroscopic variations always affect the façade aesthetic, the microscopic changes can be visible or not depending on the device itself.

This general introduction is completed by the following subsections which describe in detail the functioning of the most promising and developed technologies—grouped according to their functioning—to provide an in-depth overview of the potential and versatility of these technologies. Considering the aim of this book,

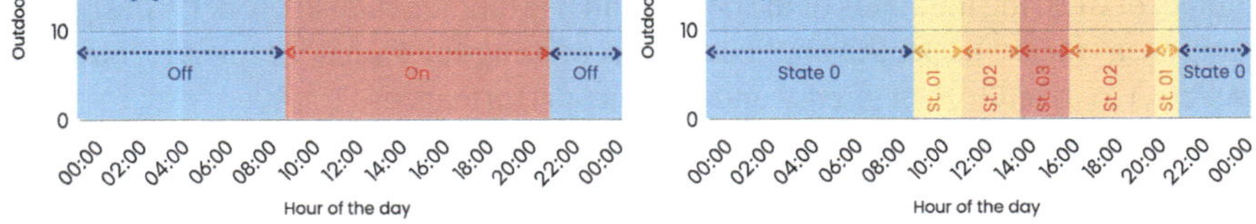

Fig. 3.2 Differences between a two-states (on–off) responsive device **a)** and a gradually adjustable technology **b)** controlled by outdoor air temperature

the described responsive systems are related to the thermal comfort and energy efficiency fields while other systems—e.g., those adopted for acoustic, indoor air quality, privacy applications, etc.—are not included in the following paragraphs to focus the attention on the key topic. Hence, the responsive technologies considered are grouped according to their contribution to:

- the improvement of the thermal inertia;
- the control of the solar radiation, further subdivide in two macro-categories of technologies (switchable windows and dynamic shadings);
- the control of the heat flux.

3.2.1 Improvement of the Thermal Inertia: Phase Change Materials

The above introduction foreshadows the wide range of possible responsive technologies; among these, the Phase Change Materials (PCMs) are undoubtedly one of the most developed intrinsically controlled technology thanks to a broad experimentation conducted in past years [27, 28] in different fields including the buildings energy efficiency [29], waste heat recovery devices [30], the electronic [31], and aerospace field [32]. The main advantage of these materials lies in their capability to store and release naturally available heat in low-volume elements, improving their energy storage density [33]. Focusing the attention on the building design domain, the PCMs can mainly bring benefits to two macro topics investigated: the reduction of thermal consumptions and the reduction of thermal discomfort. Specific studies [34] outlined the contribution of PCMs to the reduction of HVAC energy demand which could be decreased by up to 30% when applied as retrofit solutions for residential buildings in cold climates. Furthermore, these materials can also significantly reduce the thermal discomfort by stabilising the indoor radiant temperature [35].

These advantages are strictly related to the physical behaviour of the material; hence, it is fundamental to understand their nature and functioning to properly design a PCM-integrated building element. Every material can store and release heat through three different approaches: sensible heat storage (SHS), latent heat storage (LHS), and thermochemical heat storage. The latter is usually not applied to the building field while SHS and LHS strongly contribute to the thermal behaviour of the building [33]. Considering the SHS, the heat absorbed or released is related to the increase/decrease of temperature (sensible heat, Q_S) in relation to the mass of the body (m), the specific heat (c), and the variation of temperature (dT), as described in Eq. (3.1).

$$Q_S = \int_{T1}^{T2} mcdT \tag{3.1}$$

On the contrary, during the LHS, the temperature of the material does not change as the heat absorbed/released triggers the change of phase as it happens, for example, in the melting from solid to liquid. In the LHS phase, the latent heat stored (Q_L) depends on the mass, the fraction melted (f_m), and the variation of enthalpy of fusion per unit mass (Δh_m) (Eq. 3.2).

$$Q_L = mf_m\Delta h_m \tag{3.2}$$

Usually, both SHS and LHS can take part to thermal phenomena; for example, considering a gypsum-PCM board—one of the most common commercial PCM product in the building field as better described below—, the exchanged heat corresponds to the enthalpy variation of each material embedded in the board according to Eqs. (3.3), (3.4), and (3.5) [36]:

$$\Delta H_{tot} = m \cdot [(1-f)\cdot\Delta h_{gypsum} + f \cdot \Delta h_{PCM}] \tag{3.3}$$

where, considering the complete melting of the PCM ($f_m = 1$), the partial enthalpies are:

$$\Delta h_{gypsum} = \int_{T1}^{T2} mc_S dT \tag{3.4}$$

$$\Delta h_{PCM} = \left(\int_{T1}^{Tm} c_S dT + \Delta h_m + \int_{Tm}^{T2} c_L dT\right) \tag{3.5}$$

where f is the mass fraction of the PCMs in gypsum, c_S and c_L are, respectively, the specific heat of the solid and liquid states, and Tm is the melting temperature of the PCM. Figure 3.3 shows the benefits of embedding a PCM in a gypsum plasterboard in a heat-temperature diagram; thanks to its capacity to store latent heat during the melting phase, the embedded PCM avoids the increase of the surface temperature registered in a traditional plasterboard.

Considering their functioning, it is clear that PCMs cannot be extrinsically controlled due to the nature of the thermophysical reactions triggered by the rising/lowering temperature. It follows that, in the building field, to take advantage from the LHS, the melting temperature of the PCMs should be within the human comfort range (20–30 °C) [37]. In these cases, these materials can store a good amount of heat in low-volume elements without increasing their temperature improving the users' thermal comfort. From a theoretical point of view, PCMs can store or release heat in different phase transitions: solid–solid, solid–liquid, gas–liquid, gas–solid [38]; nevertheless, usually, only the solid–liquid transitions take place in the building field.

Depending on their composition, PCMs can be classified in three main classes: organic, inorganic, and eutectic. The first ones are composed of paraffins, fatty acids, fatty-acid esters, and sugar alcohols and, in general, can be classified as paraffin or non-paraffin. The main strength of organic PCMs is that they are not affected by phase segregation during repeated melting–freezing cycles and that they are slightly affected by supercooling (i.e., the PCM do not solidify below the solidification temperature). On the contrary, as paraffinic PCMs are derived from oil

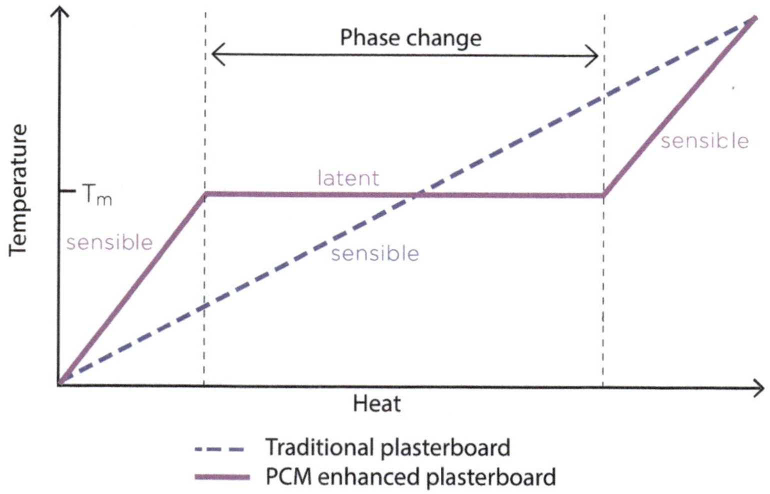

Fig. 3.3 Differences between a traditional plasterboard and a PCM enhanced plasterboard

refining, their main weakness is the low ignition resistance that makes difficult the use of these materials for envelope applications. Considering the melting temperature, organic PCMs ranges from −57 °C to +187 °C with a melting latent heat that varies between 85 and 300 J/g [39–41]; it follows that the choice of the right PCM is fundamental to optimise their functioning.

Inorganic PCMs can be classified as salt-hydrates or metallic and the main difference with the organic PCMs is that they have higher density that allows to reach higher melting latent heat per unit volume, notwithstanding the similar enthalpy per mass. Compared with organic PCMs, inorganic ones are characterised by higher ignition resistance, higher conductivity, and higher melting points. The main disadvantages of these materials are related to the risk of supercooling, phase segregation, and corrosion. Finally, the eutectic PCMs combine the main properties of different kind of PCMs with the same melting points; depending on the materials adopted to create the eutectic PCMs, they can be classified as organic–organic, inorganic–inorganic, and organic–inorganic.

From a technological point of view, to avoid damaging interactions between PCMs and the environment or other materials, they are usually encapsulated in capsules made of different materials—aluminium, plastic, polyolefin, rubber, polymers, etc.—and different shapes (balls, tubes, plates, and boxes) [42]. This process can also help to reduce the risk of corrosion and phase segregation improving, besides, their handling and the heat exchange surface [42, 43]. According to the size of the adopted capsules, the encapsulation process can be classified as macroencapsulation (d > 1 mm), microencapsulation (1 μm < d < 1 mm), or nanoencapsulation (d < 1 μm). From a technological point of view, the application of PCMs in the building field was largely studied and most of the solutions developed can be classified into two main categories: active and passive storage systems. The formers work using heat exchangers and forced convection and are classified into direct and indirect systems. In indirect systems the transfer medium (the fluid) and the storage element are distinguished; on the contrary in direct systems the heat transfer fluid performs both functions [44]. Many applications were developed for active systems [45] applied to different building elements such as suspended ceilings [46, 47], ventilation systems [48, 49], double-skin envelopes [50], solar collectors [51, 52], heat storage water tanks [53, 54], integrated photovoltaics [55, 56], and building cores enhanced with PCMs activated through the use of pipes [57].

On the contrary, no forced convection can be used in passive systems; hence, these systems are simply classified depending on the placement of the PCMs in the building element, thus: inside the material, as a new layer, and in windows or as sun protection [58]. When encapsulated, PCMs can be easily embedded in other construction materials (concrete [59], plaster [60], cellulose, or glass fibres [61]) increasing the overall thermal inertia of the building component. Another way to increase the thermal inertia—usually with reference to lightweight constructions—is to create a new layer with PCMs embedded boards; in particular, PCMs-enhanced gypsum plasterboards or PCM sandwich panels [62] are often used as additional layers to improve the envelope overall performance. The use of PCMs as additional layers is quite widespread thanks to the different commercial products already available. For example, among the microencapsulated PCMs-enhanced gypsum plasterboards Rigips-Saint Gobain markets the Alba Balance board, while Delta-Cool24 (Dorken) or the Energain (Dupont) are other commercial products that uses macroencapsulated PCMs in plates or bags.

As already mentioned, the broad variability of the PCMs melting point requires a dedicated study to select the proper material for each application. In accordance with previous studies [38, 63–65] the optimal melting point for cooling applications is up to 21 °C, while for heating applications the optimal melting point is 22 °C or 2 °C above the heating setpoint. Regarding other applications of PCMs in the building field, the optimal melting temperature for thermal comfort ranges between 22 and 28 °C while hot water applications require higher temperatures usually between 29 and 60 °C. Instead, very high melting points are required for waste heat recovery applications.

To sum up, when applied to building envelopes, PCMs can improve the time shift of the summer heat loads, increase the thermal mass and the heat storage in low-volume elements, reduce the surface temperature increasing the thermal comfort, and do not require specific maintenance. On the other hand, the paraffine flammability significantly reduces the fire safety of many of these technologies; moreover, the choice of the right PCMs is strictly climate dependant and do not allow optimisation or fine tuning during the component service life due to its intrinsic control.

PCM windows

A specific less common application of PCMs in the building field regards the PCM windows prototypes which aims to improve the inertia of transparent envelopes. Indeed, thanks to the spread of glazed façade in the last years, the thermal, solar, and visible performances were significantly improved, while the low thermal inertia of windows is still an open topic. Considering their functioning, this technology can be classified also as an intrinsically controlled switchable glazing; nevertheless, it has been included in this section according to its application field and physical functioning. The PCM windows try to increase the thermal inertia of glazed envelopes through the use of translucent PCMs within the windows' cavity. These systems are usually based on paraffin PCMs which are translucent in their solid state and fully transparent in the liquid state. An example of paraffine wax PCM windows was proposed by Goia et al. [66] who studied a 8–15-6 mm Double Glazing Unit (DGU) with a cavity filled with a Rubitherm commercial grade paraffine wax (melting temperature of 35° C and a heat store capacity of 170 J/g). The experimental results highlighted that the increased thermal inertia led to an improved energy performance during summer in a temperate climate. Another study [67] evaluated the effect of the use of PCM windows from a daylighting point of view; in particular, in this study, the PCM windows showed a better visual comfort (intended as useful illuminance and Daylight Glare Probability) only when sky luminance is low. The main issue of PCM windows is that the melting and solidification act simultaneously on different physical domains reducing the control of the system. While, on the one hand, certain behaviours such as the heat storage during the melting phase are improved, on the other hand, the summer solar properties are worsened in the liquid phase, hence, predicting their behaviour could be more complex. Despite this, the benefits on the thermal inertia can be fundamental to improve the overall windows performance, therefore it could be worth further considering this technology. Considering that this is a specific application of the PCMs, the same abovementioned advantages and disadvantages of opaque PCMs can be identified. Additionally, on the one hand, the scattering effect of the PCM in its solid phase allows to reduce the glare risk while, on the other hand, it reduces the external view and the incoming solar radiation during winter [68, 69]. This latter disadvantage is one of the greatest limits of this technology as improvement in the summer period corresponds to worsening during winter from a solar radiation point of view.

3.2.2 Solar Radiation Control: Switchable Windows

Unlike the PCMs, the switchable windows considered in this section act directly on the control of the solar radiation to improve the energy performance of the building by changing their properties between two or more states according to their composition. Considering the recent spread of transparent envelopes, these systems are catching the attention of researchers and manufacturers thanks to their capability to maximise the solar gains during winter and reduce the solar loads during summer. Moreover, thanks to their suitability for retrofitting interventions, they are set to be one of the most promising responsive technologies. Each of the following subsections describes in detail the most diffuse and promising technologies—classified according to their operation—adopted in the building field domain excluding all those devices (such as Liquid Crystal, Suspended Particle Devices, etc.) adopted only for aesthetic or privacy applications or in other application fields [70].

3.2.2.1 Extrinsically Controlled Glazing

Electrochromic, plasmonic electrochromic, and nanocrystal in-glass composites

One of the most evolved active switchable systems is, undoubtedly, the Electrochromic (EC) technology thanks to its capability to switch—according to customisable settings—its solar and optical transmittance properties. The switch between bleached and coloured state is allowed thanks to a reversible oxidation or reduction reaction that occurs as a response to an electrical stimulus in a 5-layers coating applied to the glazed panes. The coating outer layers are two external transparent conductive layers respectively followed by EC films deposited on each conductor and an intermediate electrolyte layer [71, 72] as shown in Fig. 3.4.

When an external low voltage is applied to the transparent conductive layers, the cations—usually H^+ or Li^+—embedded in the electrolyte are displaced from the cathodic to the anodic or ion-storage coating. Hence, due to this lack of electrical balance, electrons move from the conductors to the coatings, or vice versa, changing the transmittance properties of the window. EC devices can be classified as Conventional ElectroChromic (CEC), Near Infrared Radiation switching EC (NEC), and Dual-Band ElectroChromic (DBEC) according to the specific EC film adopted which can act on specific solar radiation wavelengths. While CECs can change their transmittance properties contemporaneously on Near Infrared Radiation (NIR) and on Visible Light (VL) spectra, NECs act only on the NIR spectrum changing the thermal behaviour without affecting the daylighting. A more flexible behaviour is offered by the DBECs which can assume three different states: transparent in both

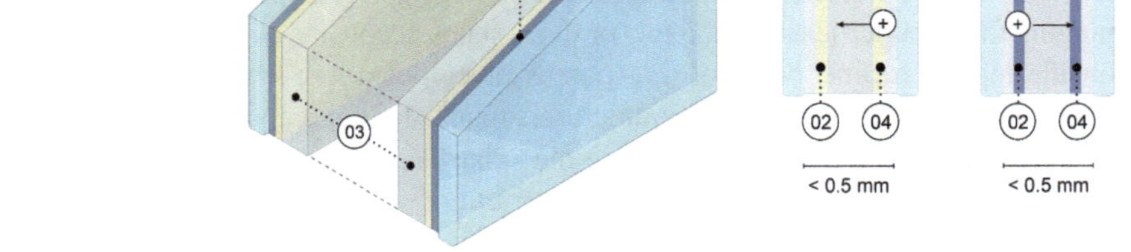

Fig. 3.4 Typical structure and functioning of an EC device [74]

NIR and VL spectra, dark in the NIR spectrum and transparent in the VL spectrum, and dark in both NIR and VL spectra [73].

In the NECs the use of doped semiconducting nanocrystals allows the modulation of the surface plasmon [75] that increases the spectral shift. For example, using the tin doped indium oxide (ITO)—which is characterised by a localised surface plasmon resonance (LSPR) in the NIR spectrum—allows to reach a larger spectral shift thanks to the electrochemical doping effect [76]. The DBECs are based on a similar functioning but their different structure results, sometimes, in a different definition (nanocrystal in-glass composites); despite this, as the functioning criterion is exactly the same of the CEC, they can be considered simply a different EC application. This technology uses the ITO nanocrystals embedded in a niobium oxide (NbO_x) glassy matrix [77]; this specific structure allows to change the optical properties in different ways according to the applied voltage as the nanocrystals and the glassy matrix perform differently. When the highest voltage is applied (4 V) both materials are in their clear configuration, when the voltage is decreased (2.3 V) only the nanocrystals—which act on the NIR spectrum—turn dark and, finally, with the lowest voltage (1.5 V) the device acts on both NIR and VL spectra as both materials are in their dark states [77].

To sum up, among the main advantages of EC technology, we can consider [69, 78, 79]:

- possible complex sensing and control algorithms to improve energy efficiency and IEQ;
- low switching voltage (1–5 V);
- good open circuit or optical memory;
- power absorption only during the switching phase;
- the possibility to apply to large areas;
- a good building integration and retrofitting suitability;
- low maintenance costs.

On the other hand, the high investment cost and the aesthetical issues related to their chromatic change could slow down their spread [69].

Colour-temperature-tunable window

As just mentioned, one of the limits of the EC devices is the aesthetical issue related to their changing colour. The colour-temperature-tunable windows try to overcome this issue thanks to an additional feature that allows the control of the light colour temperature and not only the modulation of the quantity of light transmitted (used for shading purposes). This property is obtained through the use of an electrophoretic dispersion of dual-particles of 2 biprimary complementary colours controlled by three properly located electrodes that move the coloured particles in the electrophoretic dispersion. This displacement leads to different states of the window from the neutral clear state to the neutral dark one [80].

Gasochromic glazing

The gasochromic glazing (GC) is an extrinsically controlled technology based on the use of a thin gasochromic layer deposited on the glass pane. This GC layer reacts with certain chemicals, such as H_2 and O_2, that trigger a reversible coloration of the layer changing the overall transmittance properties of the window [81]. Unlike the EC systems, the GC devices are less complex and expensive as they are based on a single GC porous layer (Fig. 3.5) instead of a 5 layers EC system. From a chemical point of view, the GC layer is made of a tungsten oxide (WO_3) film with different textures, morphologies, and compositions even though recent studies [82] are also considering the molybdenum oxide (MoO_3), which is usually adopted as photochromic film, as GC film. In order to trigger the chemical reactions, the window cavity is filled with H_2 or O_2 to switch, respectively, from bleached to dark state or from dark to bleached state [83]. The window cavity—which is usually filled with Argon to stabilise H_2 and reduce the thermal transmittance—is connected to a piping and pumping system that can fill the cavity with these gases according to the required state. The switch from clear to dark state is due to a chemical reaction on the GC layer—which is covered by a thin (4–5 nm) platinum or palladium catalyst—triggered by the H_2 pumped in the cavity. When the cavity is filled with O_2, the chemical bond between H_2 and O_2 (that produces H_2O) allows to return to the clear configuration as shown in Fig. 3.5.

Among the main advantages of GC devices we can mention the lower cost of the device compared with EC systems thanks to its simple layer configuration [84], possible complex sensing and control algorithms. Nevertheless, the starting enthusiasm regarding this technology has gradually faded because its effect is mainly limited to the visible part of the spectrum [85] and because of the maintenance of the piping system.

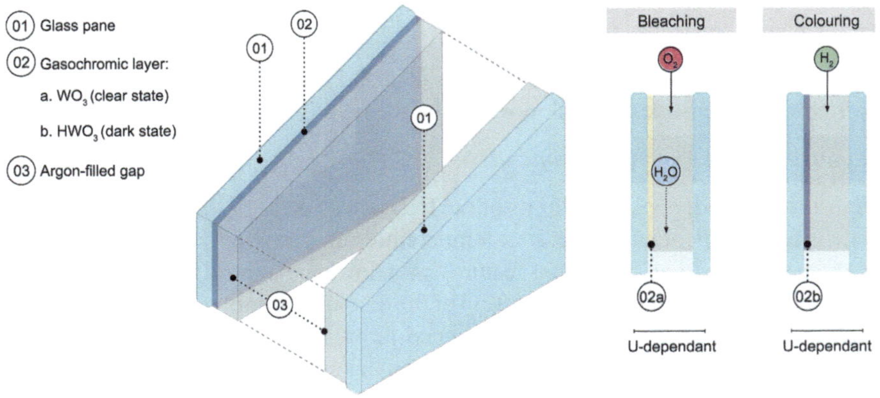

Fig. 3.5 Typical structure and functioning—schematic and not to scale—of WO3 GC devices [74]

Deformation tunable device

While EC and GC windows can be considered similar from a technological point of view, the so-called Deformation Tunable Devices are based on a completely different functioning. In particular, rather than considering a chemical reaction, the Deformation Tunable Devices change their properties through a microscale deformation of the device itself. The windows opacity can be changed wrinkling the window surface through reversible microscopic geometric deformations that change the light scattering. Compared with chromogenic materials, the main advantage of this technology relies in its colour neutrality in all the configurations. Controlling the surface topography is a developing technique which is going to increase its spread thanks to the wide experimentations conducted in different fields such as dry adhesives or micro-lens arrays [86].

These devices are composed by silver nanowires embedded in a soft dielectric elastomer; when a specific voltage is applied to the nanowires, the elastomer changes its geometry from a microscopic point of view. It follows that this new irregular surface changes the light refraction with a resulting lower transmittance at all the wavelengths [87].

Considering the same functioning criterion, the deformation of the surface can be also obtained triggering the wrinkling or flattening of the surface directly applying the strain to specific devices such as those composed by silica particles embedded in a polydimethylsiloxane (PDMS) film [88]. Considering these specific systems, the activating input is the mechanical stretching or release input rather than the electrical voltage variation. Nevertheless, despite the potential of these systems, currently no specific studies applied to the energy behaviour of buildings are available.

3.2.2.2 Intrinsically Controlled Glazing

Thermochromic windows

Moving towards intrinsically controlled systems, Thermochromic (TC) windows are one of the most developed systems thanks to their passive behaviour driven by the temperature variation. TC materials can dynamically change their properties in accordance with their temperature triggering a self-adapting behaviour as response to the external environmental variations. Each TC material is characterised by a certain critical transition temperature. Below this threshold the material is in its monoclinic state and acts as a semiconductor, less reflective in the NIR spectrum. On the contrary, exceeding the transition temperature, the TC material changes its internal configuration changing from monoclinic to rutile state; in the rutile state the material behaves like a semi-metal increasing, hence, the NIR reflections [89].

It follows that these systems do not require any sensors, controllers, or actuators simplifying significantly the overall complexity of these devices. Among the TC materials, the most adopted for windows applications are the vanadium dioxide (VO_2) [90], hydrogels, liquid crystals, ionic liquids [91], and perovskite [92]. Two macro behaviour can be identified in these systems; in particular, during the crystal transition VO_2 nanocrystals, ionic liquids, and perovskite TC materials can directly change the absorbance intensity or the absorbance peak. A slightly different behaviour characterises the hydrogels and liquid crystals which vary their scattering and reflecting properties acting on the phase separation and on the orientation of the crystals rather than on the spectral absorbance [93]. The most considered TC material for windows application is the VO_2 which is characterised by a transition temperature of 68 °C that triggers phase transition of the crystals. As previously mentioned, the transition in these materials improves the absorbance in the NIR spectrum without affecting the VL and UV wavelengths. Clearly, for building applications, this behaviour may need some improvements for example to reduce the transition temperature or to act also on the VL spectrum. To that end, specific dopants such as Tungsten and Magnesium can be added respectively to reduce the transition temperature and to change the VL absorbance. Moreover, these devices can be coupled with other technologies—such as electrochromic, electrothermal, or photochromic materials—to develop highly performance multifunctional TC windows [93]. Summing up the main features of these materials, their intrinsic control is undoubtedly an advantage from cost and maintenance points of view; nevertheless, it gets more difficult to predict their behaviour during the service life of the building.

Photochromic glazing

The Photochromic (PC) systems are based on the chromogenic properties of certain PC materials that change their colour in a reversible manner when exposed to specific solar wavelengths. In the building field, the PC materials are usually embedded in a transparent matrix to create a PC film to be applied on a transparent substrate—usually glass—to add a dynamic behaviour to the windows. This dynamism can be obtained with different PC materials, classified as organic or inorganic. The former, are characterised by a transition between two chemical isomers triggered by solar radiation; this transition leads to a change in the absorbance spectra that provides the required dynamism of the solar properties. Organic PC molecules can, in turn, be classified as thermal reversible—or thermally stable—and thermal irreversible—or thermally unstable—depending on their bleaching mode. The thermally stable molecules return in their clear state when exposed to visible light; on the contrary, the thermally unstable molecules turn back to their original colour when heated or irradiated with light [93]. Molecules such as diarylethene [94], spiropyran, and spirooxazine [95] are included in the organic class; among these, spirooxazine molecules are the most used in the building field for aesthetical reasons as they turn from transparent to a blue state rather than to that red/purple colour typical of other PC materials. Instead, the inorganic molecules – such as the transition metal oxides WO_3, TiO_2, MoO_3 etc.—behave in a different way. For example, the WO_3—which is particularly suitable for building applications for its transparent/blue colours—can change colour thanks to a redox reaction triggered by photons [96]. Hence, when the PC film is irradiated, pairs of electrons and holes are formed and the reactions between WO_3, hydrogen ions, and electrons form the H_xWO_3 compound that changes the colour of the film. When the solar radiation goes below a certain threshold, the PC film returns reversibly in its clear state. The modulation range of these devices is lower than other chromogenic technologies and this feature leads to lower energy savings in the building field, even though only few studies consider the energy benefits of these devices [72]. Similarly to other intrinsically controlled switchable devices, the maintenance advantages associated with this kind of control are counterbalanced by a higher complexity in the design phase to optimise the behaviour of the system. Moreover, the instability of the material over the long term still represents a source of uncertainty.

Photoelectrochromic and photovoltachromic glazing

The PhotoElectroChromic (PEC) devices can be considered the passive adaptation of the EC systems thanks to the use of renewable energy sources to trigger the EC redox reaction. To achieve this goal, the EC films are usually coupled with photovoltaic (PV) materials—such as dye sensitized TiO_2—to self-produce the energy required for modulating the transmittance properties. It follows that these systems can be used both as intrinsically and extrinsically controlled depending on the management of the device. PECs are usually classified with regards to materials, layers' structure, and state of aggregation [97]. Considering the structure of the systems, the EC and PV materials can be located on the same substrate—in the so-called combined devices—or can be placed on different electrodes in the separated devices [98] as shown in Fig. 3.6. A further evolution of these systems was proposed by Wu et al. [99] who developed a PhotoVoltaChromic (PVC) device to increase the amount of energy produced exceeding the threshold required for the EC films activation. After the complete transition of the EC films to the dark state, the exceeding energy produced can be used in any other electrical applications within the electrical grid. Thanks to this feature, the PVC devices can produce energy, reduce the energy demand [100], and reduce the glare discomfort.

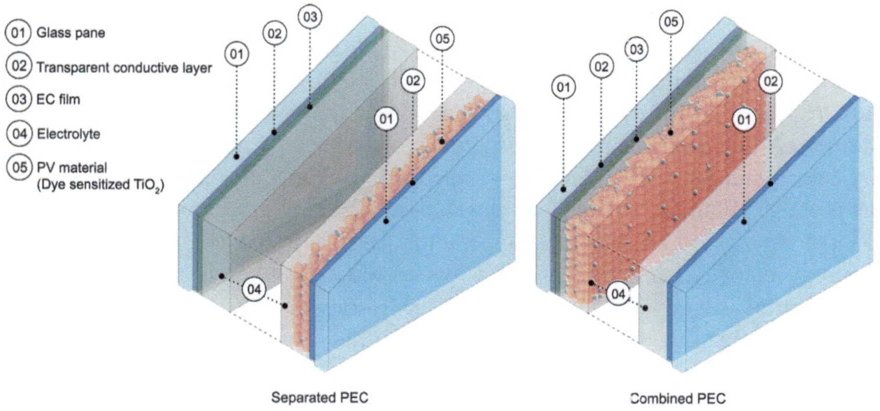

Fig. 3.6 Typical structures—schematic and not to scale—of a TiO_2 PEC device [74]

Shading fluid windows

Among the intrinsically controlled devices, a very simple and cheap system was developed by Fazel et al. [101] thanks to the shading effect of a thin coloured fluid layer. Depending on the temperature, this coloured fluid moves within a small cavity, changing the shading effect of the device. With reference to the structure of the device, the shading fluid windows can be classified as gas–liquid or liquid–liquid devices. In the former the expansion of a transparent gas pushes the shading fluid in a small cavity; on the contrary, in the liquid–liquid devices, a transparent liquid or a coloured liquid—characterised by different thermal expansion coefficients—are pushed alternatively according to the temperature variation. Despite the simplicity of this system, after the first prototypes it has not been further developed by the authors.

3.2.3 Solar Radiation Control: Dynamic Shadings

The dynamic shading systems work, basically, in the same way as switchable windows as they change their configuration to improve the solar access during the heating period and minimise the solar loads during summer. While the switchable glazing is interested by molecular changes that vary the interaction with the solar radiation, the dynamic shadings are interested by macroscopic variations (movements) that allow to block/let the radiation through the device. These systems are the evolution of the fixed shading systems already adopted in the vernacular architecture to reduce the incoming radiation in hot climates. In this case, the movement of the whole shading element allows to improve the energy performance in both heating and cooling periods, making these systems useful also in temperate climates. Starting from the first examples of dynamism in the building field—such as the "Girasole villa" designed by Angelo Invernizzi in 1930 [102]—the last decades have seen the growth of dynamic devices properly supported by countless technical and technological innovations. Also these devices are usually classified as intrinsically or extrinsically controlled depending on their operation. Looking at the intrinsically controlled systems, the most diffuse technology is based on the use of solar-activated Shape Memory Alloy (SMA) wires—or, as alternatives, Shape Memory Polymers (SMPs) or Shape Memory Hybrids (SMHs)—embedded in the shading fibres [103, 104]. The SMA wires "naturally" change their shapes according to their temperature (correlated to incident solar radiation and air temperature) triggering the movement of the whole shading element. Therefore, the solar-activated SMA shading systems can be classified as intrinsically controlled as the components for perception/cognition/action are all embedded and handled by the shading itself thanks to the SMA behaviour that "naturally" changes their shape (according to their temperature) and, consequently, the shape of the whole shading. The same system can be adopted also as extrinsically controlled device if the SMAs are triggered through electrically-induced heat [105, 106]. In this case, the perception is handled by one or more sensors (daylighting, solar

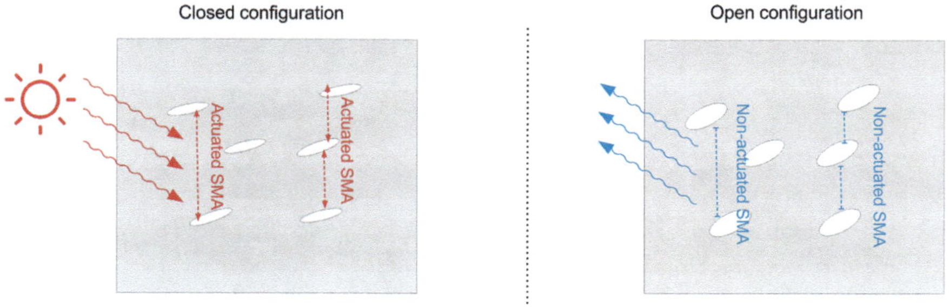

Fig. 3.7 Examples of intrinsically controlled SMA screen of the Piraeus Tower [74]

radiation, indoor/outdoor temperature, etc.), the cognition is handled by a controller or by the user, while the action is handled by the activation system.

The Piraeus Tower [103] and the Tent, Curtain and Blind prototypes [105] can be considered two examples of the different use of the SMAs in dynamic systems; the former is directly triggered by the solar radiation (Fig. 3.7) while in the latter the activating heat is produced through electrical current.

In general, the automated—or extrinsically controlled—systems can be considered more flexible and can be customised according to specific needs simply changing the activation threshold or the sensors. It follows that they are more developed and includes very heterogeneous systems ranging from SMA systems to mechanical [100] and hydraulic [107] systems. The automated mechanically activated systems are already well developed and there are different built examples ranging from very simple devices such as venetian blinds to complex three-dimensional shading patterns. The application of external mechanical stimuli triggers three different kinds of movements of the shading elements: translation, rotation (swivel, revolving, swing), or their combination (folding, expanding, and contracting) [108]. Built applications of complex automated shadings have spread since the end of last century. For example, the well-known Arab World Institute realised in Paris in 1988 by Jean Nouvel is characterised by a dynamic façade constituted by 240 main shutters that increase or reduce the amount of incoming radiation thanks to 30,000 light sensitive diaphragms (sensors), a central computer (controller), and a mechanical system (actuator) [102]. Another recent and different example of an extrinsically controlled dynamic shading is the one realised by Aedas and Arup in 2013 for the Al Bahr Towers in Dubai were the designers merged the concept of the Islamic "Mashrabiya" screens with three-dimensional mechanically activated origami umbrellas [109]. This shading system is composed by 1049 triangular origami elements that cover the glazed curtain wall of two 150 m tall circular towers on the southern, eastern, and western exposures [110].

Each element of the shading system can fold or unfold according to the sun position thanks to a barycentric pushing piston that can move the middle point of the shading element up to 1 m, folding the panels and providing up to 85% of unshaded area (Fig. 3.8) [111]. To improve the durability and the overall performances, the shading elements are made of Polytetrafluoroethylene (PTFE) coated glass fibre

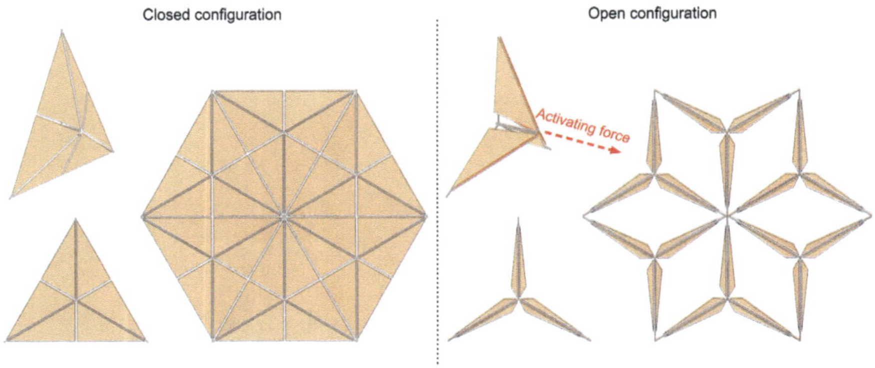

Fig. 3.8 Functioning criterion of the Al Bahr Tower shading system [74]

mesh and the system was largely tested before the realization. In particular, more than 30,000 opening-closing cycles were conducted in a climatic chamber to evaluate the behaviour under different boundary conditions such as different temperatures, presence of sand and salt water on the joints, etc.

A further boost to the development of these systems was provided by the spread of biomimetics and phytomimetics concepts [112]. Indeed, starting from nature-inspired solutions and supported by new technological advancements, a new class of shading systems based on specific fibres anisotropy was developed by researchers to improve the buildings' performance. Similarly to what happens in nastic structures, adopting specific distribution and orientation of the material's fibres allows the movement of the shading systems. This movement can be triggered directly by environmental stimuli (e.g., solar activated SMAs) in the autonomous systems or by external energy source in the automated systems. For example, Flectofin® [100] is an automated flapping shading system inspired by the Strelizia reginae plant based on the anisotropy of a Glass Fibre-Reinforced Polymer. In this device, the external compression or bending force causes the buckling of the shading element that rotates around the vertical axis reducing the incoming solar radiation.

This brief introduction to the main different available dynamic systems shows an extremely faceted scenario and, according to the device, activation criterion, and location considered, dynamic shadings can reduce energy demand up to 20–34% [113]. Considering the great heterogeneity of these devices, specific pros and cons could be identified for each technology. However, on the whole, considering automated or extrinsically controlled systems, these devices allow a wide range of possible geometrical and functional solutions which can be managed by complex sensing and control algorithms to reduce energy consumptions while improving the IEQ [71]. Nevertheless, the high maintenance cost of the mechanical systems, the high architectural integration needed, and the partial reduction of the view out are still slowing the spread of these devices. On the contrary, considering the autonomous or intrinsically controlled dynamic systems, the specific strengths and weaknesses are strictly related to the specific device adopted.

3.2.4 Control of the Heat Flux: Dynamic Insulation Systems

The concept behind these devices is based on the need to adjust the envelope thermal transmittance in order to manage the heat transfer rate through the building envelope; the aim of this approach is minimising the heating consumption and reducing the overheating risk during summer in highly insulated envelopes. Indeed, despite maximizing the insulation leads to clear benefits during winter and during summer air conditioning period, the reduction of the thermal transmittance can reduce the benefits related, for example, to the summer night heat flows. Hence, the dynamic insulation devices try to find a balance between these behaviours minimising the thermal transmittance when required (i.e., during winter or during summer when the systems are on) and increasing the heat flows when needed (i.e., during summer when the systems are off and the outdoor temperature is lower than the indoor temperature). Several authors proposed many options to achieve dynamic insulation based on different mechanisms—briefly described in Fig. 3.9—adopted as main classification system in the following subparagraphs. Regardless the technology considered, these devices are usually extrinsically controlled and many different approaches were considered to manage these systems. Possible control strategies can range from the occurrence of heating/cooling demands, up to the model predictive control techniques in which the state of the device is set out based on the results of an optimisation problem run for each control timestep.

Pressurised gas devices

For all the technologies included in this class, the responsiveness of the system is based on a micro scale variation related to the gas pressure, the mean free path of the

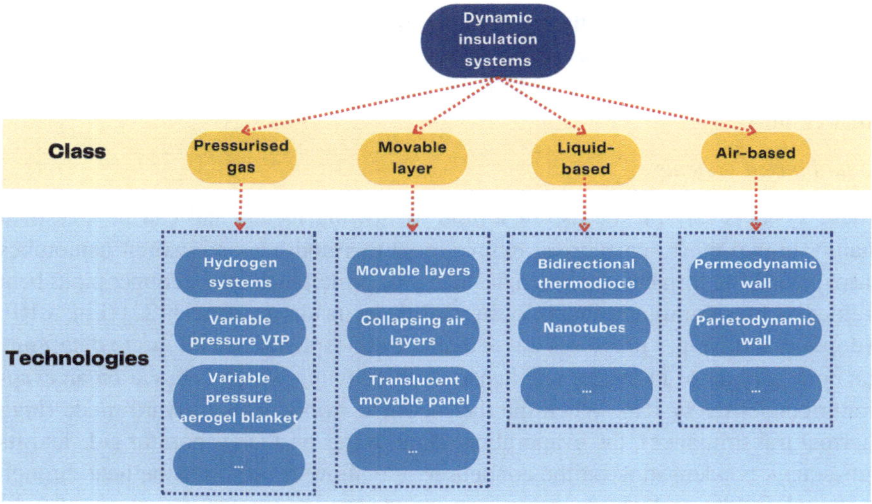

Fig. 3.9 Classification of the main dynamic insulation systems based on their mechanism

molecules, or the interaction between gas and insulated materials [114]. Xenophou patented in 1976 [115] a device which allows to change the thermal properties increasing or reducing the hydrogen pressure in the cavity of a cellular structure. Similarly, in 1994, Benson [116] based the variation of the thermal transmittance of the component on the hydrogen absorption and desorption triggered by the hydrogen pressure. A different approach was followed by Berge et al. [117] who controlled the thermal transmittance of the system changing the air pressure inside a Vacuum Insulation Panel (VIP) and in an aerogel blanket.

Movable layers

The technologies included in this class rely on a completely different concept as, basically, the variation of the thermal transmittance depends on the variation of the number or dimension of the air cavities which allow or prevent the natural convection within the envelope component. For example, the number of the envelope air cavities can be increased or decreased by using a series of rollable films that modifies the configuration of the air cavity changing, as consequence, the overall thermal transmittance as proposed by Pflug et al. [118]. In particular, rolling out the films, the heat flow through the envelope is reduced thanks to a series of thin air cavities; on the contrary, rolling up the films allows to increase the heat flow thanks to a single larger air cavity.

The same authors proposed a transparent system [119, 120] thanks to the use of a translucent movable insulating panel which can be adopted for variable U windows. Indeed, using materials such as open pore melamine foam ($\lambda = 0.035$ W/mK) or aerogel ($\lambda = 0.013$ W/mK), can provide both low conductivity and a certain amount of incoming light fundamental for these applications. The functioning is very simple as when the panel is in its top position, the air convection around it is not allowed; on the contrary, moving the panel in its middle position, a large-scale convection phenomenon is triggered by the temperature differences [121]. Despite these systems can be classified also as switchable glazing, they are included in this subsection according to their application field which is related to the heat flux rather than to the solar control.

Liquid based systems

These systems rely on the use of a fluid as flowing means and can be classified mainly in two macro categories: bidirectional thermodiode and carbon nantotubes suspension. The former is based on the heat pipe principle thanks 9 copper pipes bent welded on two aluminium sheets located inside an insulating panel [122] (Fig. 3.10). Adopting in the pipes pressures lower than the atmospheric one allows to determine the boiling point of the water and hence its functioning which is based on an evaporator/condenser system. When the thermodiode works in its forward mode (high thermal transmittance), the evaporation of the water on the evaporator side and its subsequent condensation on the condenser side allows to transfer the heat through the panel. On the contrary, in its backward mode (low thermal transmittance) thanks to the "z" shape of the pipes there is no condensation on the external surface and

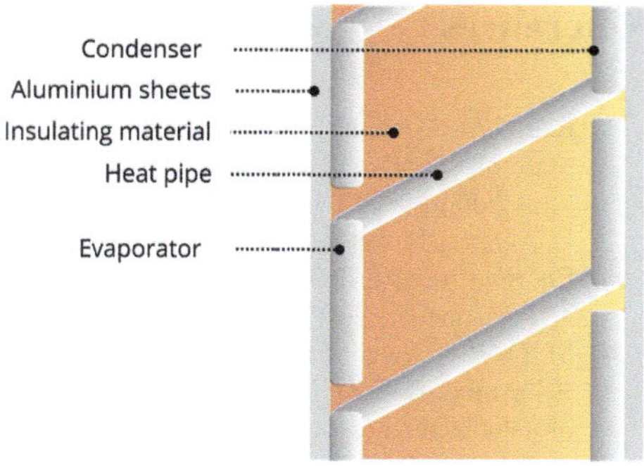

Fig. 3.10 Typical section of a thermodiode device

the heat transfer is limited only to a limited convection inside the pipes and to the thermal bridges of the copper pipes.

A completely different functioning is adopted in the carbon nanotubes suspension in a fluid [123]; in these devices the conductivity changes according to the direction of the carbon nanotubes which can be controlled varying the temperature of the fluid or its speed within the piping system.

Air based systems

Finally, air-flowing devices can be classified as parietodynamic, if the flow is parallel to the façade plane, or permeodynamic systems, if the flow is perpendicular to the wall plane. These systems can be seen as heat exchangers and the flowing air can be preheated during winter or precooled during summer; moreover, they can work pro-flux or contra-flux—depending on the relative directions of the airflow and heat fluxes—allowing to control the heat flow during different seasons [124].

Parietodynamic systems use a series of air cavities in the building walls to circulate air; basically, the wall is a heat exchanger and the air flow parallel to the wall triggers the heat flow. In this case, the air flow is limited to one or more channels coplanar with the envelope surface. On the contrary, permeodynamic systems are basically breathing walls as the air flow is perpendicular to the façade. The core of a permeodynamic device is its porous breathing layer—which is located within the wall—that allows the air flow triggered by the pressure difference when the system is in its conductive state. Moreover, these systems can improve the indoor air quality thanks to the elimination of the pollution due to the filtration through the porous material. Nevertheless, the main weaknesses of these devices are connected to possible structural damages related to wrong installation, difficult design [124], and high costs related to the air tightness of the systems [114].

References

1. R.C.G.M. Loonen, M. Trčka, D. Cóstola, J.L.M. Hensen, Climate adaptive building shells: State-of-the-art and future challenges. Renew. Sustain. Energy Rev. **25**, 483–493 (2013). https://doi.org/10.1016/j.rser.2013.04.016
2. D. Aelenei, L. Aelenei, C.P. Vieira, Adaptive façade: concept, applications, research questions. Energy Proc. **91**, 269–275 (2016). https://doi.org/10.1016/j.egypro.2016.06.218
3. N. Negroponte, Soft architecture machines, The New Media Reader (1975)
4. M. Davies, A wall for all seasons. RIBA J. **88**, 55–57 (1981)
5. F. Favoino, F. Goia, M. Perino, V. Serra, Experimental assessment of the energy performance of an advanced responsive multifunctional façade module. Energy Build. (2014). https://doi.org/10.1016/j.enbuild.2013.08.066
6. S. Attia, S. Bilir, T. Safy, C. Struck, R. Loonen, F. Goia, Current trends and future challenges in the performance assessment of adaptive façade systems. Energy Build. **179**, 165–182 (2018). https://doi.org/10.1016/j.enbuild.2018.09.017
7. J. Moloney, A framework for the design of kinetic façades, in *Computer-Aided Architectural Design Futures, CAADFutures 2007—Proceedings of the 12th International CAADFutures Conference* (2007). https://doi.org/10.1007/978-1-4020-6528-6_34
8. M. Konstantoglou, A. Tsangrassoulis, Dynamic operation of daylighting and shading systems: a literature review. Renew. Sustain. Energy Rev. (2016). https://doi.org/10.1016/j.rser.2015.12.246
9. M. Wigginton, *Intelligent Skins* (2013). https://doi.org/10.4324/9780080495446
10. F. Favoino, L. Giovannini, R. Loonen, Smart glazing in Intelligent Buildings: what can we simulate? in *GPD 2017—Glass Performance Days 2017*, pp. 8–15 (2017)
11. A. Ghosh, B. Norton, Advances in switchable and highly insulating autonomous (self-powered) glazing systems for adaptive low energy buildings, Renew Energy (2018). https://doi.org/10.1016/j.renene.2018.04.038
12. F. Carlucci, A review of smart and responsive building technologies and their classifications. Future Cities Environ. **7** (2021). https://doi.org/10.5334/FCE.123/METRICS/
13. P. Heiselberg, *IEA ECBCS Annex 44 Integrating Environmentally Responsive Elements in Buildings—Expert Guide Part 1: Responsive Building Concepts, Foundations* (2009)
14. A. van der Aa, P. Heiselberg, M. Perino, Designing with responsive building elements, in *IEA-ECBCS Annex 44 Integrating Environmentally Responsive Elements in Buildings* (2011).
15. R.C.G.M. Loonen, *Approaches for Computational Performance Optimization of Innovative Adaptive Façade Concepts* (Technische Universiteit Eindhoven, 2018). www.tue.nl/taverne
16. J. Wang, L.O. Beltrán, J. Kim, From static to kinetic: A review of acclimated kinetic building envelopes, in *World Renewable Energy Forum, WREF 2012, Including World Renewable Energy Congress XII and Colorado Renewable Energy Society (CRES) Annual Conference* (2012)
17. B. Hayes-Roth, An architecture for adaptive intelligent systems. Artif. Intell. (1995). https://doi.org/10.1016/0004-3702(94)00004-K
18. R. Romano, L. Aelenei, D. Aelenei, E.S. Mazzucchelli, What is an adaptive façade? Analysis of recent terms and definitions from an international perspective. J. Facade Design Eng. **6**, 065–076 (2018). https://doi.org/10.7480/jfde.2018.3.2478
19. D.M. Addington, D. L.shock, *Smart Materials and New Technologies For the Architecture and Design Professions* (Elsevier, 2005)
20. T.M. Lawrence, M.C. Boudreau, L. Helsen, G. Henze, J. Mohammadpour, D. Noonan, D. Patteeuw, S. Pless, R.T. Watson, Ten questions concerning integrating smart buildings into the smart grid. Build. Environ. (2016). https://doi.org/10.1016/j.buildenv.2016.08.022
21. B.L.H. Hasselaar, Climate adaptive skins: towards the new energy-efficient façade. WIT Trans. Ecol. Environ. **99**, 351–360 (2006). https://doi.org/10.2495/RAV060351
22. S. Attia, R. Lioure, Q. Declaude, Future trends and main concepts of adaptive facade systems. Energy Sci. Eng. (2020). https://doi.org/10.1002/ese3.725

23. R. Loonen, J. Rico-Martinez, F. Favoino, M. Brzezicki, C. Menezo, G. La Ferla, L. Aelenei, Design for façade adaptability: towards a unified and systematic characterization, in *10th Conference on Advanced Building Skins* (2015)
24. N. Ramzy, H. Fayed, Kinetic systems in architecture: new approach for environmental control systems and context-sensitive buildings. Sustain. Cities Soc. (2011). https://doi.org/10.1016/j.scs.2011.07.004
25. C.E. Ochoa, I.G. Capeluto, Strategic decision-making for intelligent buildings: comparative impact of passive design strategies and active features in a hot climate. Build. Environ. (2008). https://doi.org/10.1016/j.buildenv.2007.10.018
26. P.V. Rathod, J.M.C. Puguan, H. Kim, Thermo- and electrochromic smart window derived from a viologen-tethered triazolum based poly(NIPAmn-TEG-BPV) electrolyte to enhance building energy efficiency and visual comfort. Chem. Eng. J. **455**, 140874 (2023). https://doi.org/10.1016/J.CEJ.2022.140874
27. F. Kuznik, D. David, K. Johannes, J.J. Roux, A review on phase change materials integrated in building walls. Renew. Sustain. Energy Rev. (2011). https://doi.org/10.1016/j.rser.2010.08.019
28. C.N. Elias, V.N. Stathopoulos, A comprehensive review of recent advances in materials aspects of phase change materials in thermal energy storage. Energy Proc. **161**, 385–394 (2019). https://doi.org/10.1016/j.egypro.2019.02.101
29. N. Soares, J.J. Costa, A.R. Gaspar, P. Santos, Review of passive PCM latent heat thermal energy storage systems towards buildings' energy efficiency. Energy Build. (2013). https://doi.org/10.1016/j.enbuild.2012.12.042
30. H. Akeiber, P. Nejat, M.Z.A. Majid, M.A. Wahid, F. Jomehzadeh, I. Zeynali Famileh, J.K. Calautit, B.R. Hughes, S.A. Zaki, A review on phase change material (PCM) for sustainable passive cooling in building envelopes. Renew. Sustain. Energy Rev. **60**, 1470–1497 (2016). https://doi.org/10.1016/j.rser.2016.03.036
31. A. Farzanehnia, M. Khatibi, M. Sardarabadi, M. Passandideh-Fard, Experimental investigation of multiwall carbon nanotube/paraffin based heat sink for electronic device thermal management. Energy Convers. Manag. (2019). https://doi.org/10.1016/j.enconman.2018.10.037
32. G. Neri, A. Koehler, M.N. De Parolis, V. Zolesi, ESA conditioned container: a system for passive temperature controlled transportation of experiments for the international space station, in *Proceedings of the International Astronautical Congress*, IAC, Naples, Italy (n.d.)
33. A. De Gracia, L.F. Cabeza, Phase change materials and thermal energy storage for buildings. Energy Build. (2015). https://doi.org/10.1016/j.enbuild.2015.06.007
34. U. Berardi, S. Soudian, Benefits of latent thermal energy storage in the retrofit of Canadian high-rise residential buildings. Build. Simul. (2018). https://doi.org/10.1007/s12273-018-0436-x
35. F. Fiorito, Phase-change materials for indoor comfort improvement in lightweight buildings. A parametric analysis for Australian climates. Energy Proc. **57**, 2014–2022 (2014). https://doi.org/10.1016/j.egypro.2014.10.066
36. V. De Matteis, A. Cannavale, F. Martellotta, R. Rinaldi, P. Calcagnile, F. Ferrari, U. Ayr, F. Fiorito, Nano-encapsulation of phase change materials: from design to thermal performance, simulations and toxicological assessment. Energy Build. **188–189**, 1–11 (2019). https://doi.org/10.1016/j.enbuild.2019.02.004
37. J. Kośny, *PCM-Enhanced Building Components—An Application of Phase Change Materials in Building Envelopes and Internal Structures* (Springer, 2015). https://doi.org/10.1007/978-3-319-14286-9
38. L.F. Cabeza, A. Castell, C. Barreneche, A. De Gracia, A.I. Fernández, Materials used as PCM in thermal energy storage in buildings: a review. Renew. Sustain. Energy Rev. **15**, 1675–1695 (2011). https://doi.org/10.1016/j.rser.2010.11.018
39. M.M. Kenisarin, Thermophysical properties of some organic phase change materials for latent heat storage. A review. Solar Energy (2014). https://doi.org/10.1016/j.solener.2014.05.001

40. Y. Yuan, N. Zhang, W. Tao, X. Cao, Y. He, Fatty acids as phase change materials: a review. Renew. Sustain. Energy Rev. (2014). https://doi.org/10.1016/j.rser.2013.08.107
41. W. Su, J. Darkwa, G. Kokogiannakis, Review of solid-liquid phase change materials and their encapsulation technologies. Renew. Sustain. Energy Rev. (2015). https://doi.org/10.1016/j.rser.2015.04.044
42. P.B. Salunkhe, P.S. Shembekar, A review on effect of phase change material encapsulation on the thermal performance of a system. Renew. Sustain. Energy Rev. **16**, 5603–5616 (2012). https://doi.org/10.1016/j.rser.2012.05.037
43. Y.E. Milián, A. Gutiérrez, M. Grágeda, S. Ushak, A review on encapsulation techniques for inorganic phase change materials and the influence on their thermophysical properties. Renew. Sustain. Energy Rev. **73**, 983–999 (2017). https://doi.org/10.1016/j.rser.2017.01.159
44. A. Gil, M. Medrano, I. Martorell, A. Lázaro, P. Dolado, B. Zalba, L.F. Cabeza, State of the art on high temperature thermal energy storage for power generation. Part 1-Concepts, materials and modellization. Renew. Sustain. Energy Rev. (2010). https://doi.org/10.1016/j.rser.2009.07.035
45. L. Navarro, A. de Gracia, S. Colclough, M. Browne, S.J. McCormack, P. Griffiths, L.F. Cabeza, Thermal energy storage in building integrated thermal systems: a review. Part 1. Active storage systems. Renew. Energy **88**, 526–547 (2016). https://doi.org/10.1016/j.renene.2015.11.040
46. H. Weinläder, W. Körner, B. Strieder, A ventilated cooling ceiling with integrated latent heat storage—monitoring results. Energy Build. (2014). https://doi.org/10.1016/j.enbuild.2014.07.013
47. J. Lizana, M. de-Borja-Torrejon, A. Barrios-Padura, T. Auer, R. Chacartegui, Passive cooling through phase change materials in buildings. A critical study of implementation alternatives. Appl. Energy (2019). https://doi.org/10.1016/j.apenergy.2019.113658
48. M. Fiorentini, P. Cooper, Z. Ma, Development and optimization of an innovative HVAC system with integrated PVT and PCM thermal storage for a net-zero energy retrofitted house. Energy Build. (2015). https://doi.org/10.1016/j.enbuild.2015.02.018
49. A. Real, V. García, L. Domenech, J. Renau, N. Montés, F. Sánchez, Improvement of a heat pump based HVAC system with PCM thermal storage for cold accumulation and heat dissipation. Energy Build. (2014). https://doi.org/10.1016/j.enbuild.2014.04.029
50. A. de Gracia, L. Navarro, A. Castell, L.F. Cabeza, Energy performance of a ventilated double skin facade with PCM under different climates. Energy Build. (2015). https://doi.org/10.1016/j.enbuild.2015.01.011
51. M.K.A. Sharif, A.A. Al-Abidi, S. Mat, K. Sopian, M.H. Ruslan, M.Y. Sulaiman, M.A.M. Rosli, Review of the application of phase change material for heating and domestic hot water systems. Renew. Sustain. Energy Rev. (2015). https://doi.org/10.1016/j.rser.2014.09.034
52. M.M.A. Khan, N.I. Ibrahim, I.M. Mahbubul, H. Muhammad. Ali, R. Saidur, F.A. Al-Sulaiman, Evaluation of solar collector designs with integrated latent heat thermal energy storage: a review. Solar Energy (2018). https://doi.org/10.1016/j.solener.2018.03.014
53. L. Navarro, C. Barreneche, A. Castell, D.A.G. Redpath, P.W. Griffiths, L.F. Cabeza, High density polyethylene spheres with PCM for domestic hot water applications: water tank and laboratory scale study. J. Energy Storage (2017). https://doi.org/10.1016/j.est.2017.07.025
54. A. Najafian, F. Haghighat, A. Moreau, Integration of PCM in domestic hot water tanks: optimization for shifting peak demand. Energy Build. (2015). https://doi.org/10.1016/j.enbuild.2015.05.036
55. A. Hasan, S.J. McCormack, M.J. Huang, B. Norton, Energy and cost saving of a photovoltaic-phase change materials (PV-PCM) system through temperature regulation and performance enhancement of photovoltaics. Energies (Basel) (2014). https://doi.org/10.3390/en7031318
56. W. Lu, Z. Liu, J.F. Flor, Y. Wu, M. Yang, Investigation on designed fins-enhanced phase change materials system for thermal management of a novel building integrated concentrating PV. Appl. Energy (2018). https://doi.org/10.1016/j.apenergy.2018.05.030
57. A. Faheem, G. Ranzi, F. Fiorito, C. Lei, A numerical study on the thermal performance of night ventilated hollow core slabs cast with micro-encapsulated PCM concrete. Energy Build. (2016). https://doi.org/10.1016/j.enbuild.2016.06.014

58. L. Navarro, A. de Gracia, D. Niall, A. Castell, M. Browne, S.J. McCormack, P. Griffiths, L.F. Cabeza, Thermal energy storage in building integrated thermal systems: a review. Part 2. Integration as passive system. Renew. Energy **85**, 1334–1356 (2016). https://doi.org/10.1016/j.renene.2015.06.064
59. L.F. Cabeza, C. Castellón, M. Nogués, M. Medrano, R. Leppers, O. Zubillaga, Use of microencapsulated PCM in concrete walls for energy savings. Energy Build. (2007). https://doi.org/10.1016/j.enbuild.2006.03.030
60. S. Ramakrishnan, X. Wang, J. Sanjayan, J. Wilson, Thermal performance assessment of phase change material integrated cementitious composites in buildings: experimental and numerical approach. Appl. Energy (2017). https://doi.org/10.1016/j.apenergy.2017.05.144
61. J. Kosny, E. Kossecka, A. Brzezinski, A. Tleoubaev, D. Yarbrough, Dynamic thermal performance analysis of fiber insulations containing bio-based phase change materials (PCMs). Energy Build. (2012). https://doi.org/10.1016/j.enbuild.2012.05.021
62. C. Castellón, M. Medrano, J. Roca, L.F. Cabeza, M.E. Navarro, A.I. Fernández, A. Lázaro, B. Zalba, Effect of microencapsulated phase change material in sandwich panels. Renew. Energy (2010). https://doi.org/10.1016/j.renene.2010.03.030
63. C. Barrenechea, H. Navarro, S. Serrano, L.F. Cabeza, A.I. Fernández, New database on phase change materials for thermal energy storage in buildings to help PCM selection. Energy Proc. (2014). https://doi.org/10.1016/j.egypro.2014.10.249
64. D. Heim, J.A. Clarke, Numerical modelling and thermal simulation of PCM-gypsum composites with ESP-r. Energy Build. 795–805. Elsevier (2004). https://doi.org/10.1016/j.enbuild.2004.01.004
65. Y. Zhang, G. Zhou, K. Lin, Q. Zhang, H. Di, Application of latent heat thermal energy storage in buildings: state-of-the-art and outlook. Build. Environ. **42**, 2197–2209 (2007). https://doi.org/10.1016/j.buildenv.2006.07.023
66. F. Goia, M. Perino, V. Serra, Experimental analysis of the energy performance of a full-scale PCM glazing prototype. Sol. Energy **100**, 217–233 (2014). https://doi.org/10.1016/j.solener.2013.12.002
67. L. Giovannini, F. Goia, V.R.M. Lo Verso, V. Serra, Phase change materials in glazing: implications on light distribution and visual comfort. Preliminary results. Energy Proc. **111**, 357–366 (2017). https://doi.org/10.1016/j.egypro.2017.03.197
68. I. Vigna, L. Bianco, F. Goia, V. Serra, Phase change materials in transparent building envelopes: a strengths, weakness, opportunities and threats (SWOT) analysis. Energies (Basel) **11** (2018). https://doi.org/10.3390/en11010111
69. F. Carlucci, F. Fiorito, Implementation of smart technologies in building envelopes: methodological approach and comparative analyses in a hot-summer Mediterranean climate, in Colloqui.AT.e 2021 Design and Construction Tradition and Innovation in the Practice of Architecture, ed. by E. Sicignano (EdicomEdizioni, 2021). ISBN 978-88-96386-62-0
70. R. Baetens, B.P. Jelle, A. Gustavsen, Properties, requirements and possibilities of smart windows for dynamic daylight and solar energy control in buildings: a state-of-the-art review. Sol. Energy Mater. Sol. Cells (2010). https://doi.org/10.1016/j.solmat.2009.08.021
71. C.G. Granqvist, I. Bayrak Pehlivan, G.A. Niklasson, Electrochromics on a roll: web-coating and lamination for smart windows. Surf. Coat. Technol. (2018). https://doi.org/10.1016/j.surfcoat.2017.08.006
72. A. Cannavale, Chromogenic technologies for energy saving. Clean Technol. **2**, 462–475 (2020). https://doi.org/10.3390/cleantechnol2040029
73. N. DeForest, A. Shehabi, S. Selkowitz, D.J. Milliron, A comparative energy analysis of three electrochromic glazing technologies in commercial and residential buildings. Appl. Energy **192**, 95–109 (2017). https://doi.org/10.1016/j.apenergy.2017.02.007
74. F. Carlucci, A review of smart and responsive building technologies and their classifications. Future Cities Environ. **7** (2021). https://doi.org/10.5334/FCE.123
75. N. DeForest, A. Shehabi, J. O'Donnell, G. Garcia, J. Greenblatt, E.S. Lee, S. Selkowitz, D.J. Milliron, United States energy and CO_2 savings potential from deployment of near-infrared electrochromic window glazings. Build. Environ. (2015). https://doi.org/10.1016/j.buildenv.2015.02.021

76. G. Garcia, R. Buonsanti, A. Llordes, E.L. Runnerstrom, A. Bergerud, D.J. Milliron, Near-infrared spectrally selective plasmonic electrochromic thin films. Adv. Opt. Mater. (2013). https://doi.org/10.1002/adom.201200051
77. A. Llordés, G. Garcia, J. Gazquez, D.J. Milliron, Tunable near-infrared and visible-light transmittance in nanocrystal-in-glass composites. Nature (2013). https://doi.org/10.1038/nature12398
78. W.J. Hee, M.A. Alghoul, B. Bakhtyar, O. Elayeb, M.A. Shameri, M.S. Alrubaih, K. Sopian, The role of window glazing on daylighting and energy saving in buildings. Renew. Sustain. Energy Rev. **42**, 323–343 (2015). https://doi.org/10.1016/j.rser.2014.09.020
79. P.M. Martin, Optical materials: smart optical materials, in *Encyclopedia of Modern Optics, Five-Volume Set* (Elsevier Inc., 2004), pp. 9–16. https://doi.org/10.1016/B0-12-369395-0/00869-1
80. S. Mukherjee, W.L. Hsieh, N. Smith, M. Goulding, J. Heikenfeld, Electrokinetic pixels with biprimary inks for color displays and color-temperature-tunable smart windows. Appl. Opt. (2015). https://doi.org/10.1364/ao.54.005603
81. M. Casini, Active dynamic windows for buildings: a review. Renew. Energy **119**, 923–934 (2018). https://doi.org/10.1016/j.renene.2017.12.049
82. F. Delalat, M. Ranjbar, H. Salamati, Blue colloidal nanoparticles of molybdenum oxide by simple anodizing method: decolorization by $PdCl_2$ and observation of in-liquid gasochromic coloration. Sol. Energy Mater. Sol. Cells (2016). https://doi.org/10.1016/j.solmat.2015.08.038
83. C.M. Lampert, Chromogenic smart materials. Mater. Today (2004). https://doi.org/10.1016/S1369-7021(04)00123-3
84. W. Feng, L. Zou, G. Gao, G. Wu, J. Shen, W. Li, Gasochromic smart window: optical and thermal properties, energy simulation and feasibility analysis. Sol. Energy Mater. Sol. Cells **144**, 316–323 (2016). https://doi.org/10.1016/J.SOLMAT.2015.09.029
85. M. Brzezicki, A systematic review of the most recent concepts in smart windows technologies with a focus on electrochromics. Sustainability (Switzerland) **13**, 9604 (2021). https://doi.org/10.3390/SU13179604/S1
86. S.G. Lee, D.Y. Lee, H.S. Lim, D.H. Lee, S. Lee, K. Cho, Switchable transparency and wetting of elastomeric smart windows. Adv. Mater. **22**, 5013–5017 (2010). https://doi.org/10.1002/adma.201002320
87. S. Shian, D.R. Clarke, Electrically tunable window device. Opt. Lett. (2016). https://doi.org/10.1364/ol.41.001289
88. H.N. Kim, D. Ge, E. Lee, S. Yang, Multistate and on-demand smart windows. Adv. Mater. **30**, 1–7 (2018). https://doi.org/10.1002/adma.201803847
89. M. Kamalisarvestani, R. Saidur, S. Mekhilef, F.S. Javadi, Performance, materials and coating technologies of thermochromic thin films on smart windows. Renew. Sustain. Energy Rev. (2013). https://doi.org/10.1016/j.rser.2013.05.038
90. M.E.A. Warwick, R. Binions, Advances in thermochromic vanadium dioxide films. J. Mater. Chem. A Mater. (2014). https://doi.org/10.1039/c3ta14124a
91. Y. Ke, C. Zhou, Y. Zhou, S. Wang, S.H. Chan, Y. Long, Emerging thermal-responsive materials and integrated techniques targeting the energy-efficient smart window application. Adv. Funct. Mater. (2018). https://doi.org/10.1002/adfm.201800113
92. Y. Zhang, C.Y. Tso, J.S. Iñigo, S. Liu, H. Miyazaki, C.Y.H. Chao, K.M. Yu, Perovskite thermochromic smart window: advanced optical properties and low transition temperature. Appl. Energy (2019). https://doi.org/10.1016/j.apenergy.2019.113690
93. Y. Ke, J. Chen, G. Lin, S. Wang, Y. Zhou, J. Yin, P.S. Lee, Y. Long, Smart windows: electro-, thermo-, mechano-, photochromics, and beyond. Adv. Energy Mater. **9**, 1–38 (2019). https://doi.org/10.1002/aenm.201902066
94. G.H. Timmermans, B.W.H. Saes, M.G. Debije, Dual-responsive "smart" window and visually attractive coating based on a diarylethene photochromic dye. Appl. Opt. (2019). https://doi.org/10.1364/ao.58.009823
95. P. Miluski, M. Kochanowicz, J. Zmojda, D. Dorosz, Optical properties of spirooxazine-doped PMMA fiber for new functional applications. Opt. Eng. **56**, 047105 (2017). https://doi.org/10.1117/1.oe.56.4.047105

96. S. Wang, W. Fan, Z. Liu, A. Yu, X. Jiang, Advances on tungsten oxide based photochromic materials: strategies to improve their photochromic properties. J. Mater. Chem. C Mater. (2018). https://doi.org/10.1039/c7tc04189f
97. A. Cannavale, P. Cossari, G.E. Eperon, S. Colella, F. Fiorito, G. Gigli, H.J. Snaith, A. Listorti, Forthcoming perspectives of photoelectrochromic devices: a critical review. Energy Environ. Sci. (2016). https://doi.org/10.1039/c6ee01514j
98. A. Cannavale, F. Martellotta, F. Fiorito, U. Ayr, The challenge for building integration of highly transparent photovoltaics and photoelectrochromic devices. Energies (Basel) **13** (2020). https://doi.org/10.3390/en13081929
99. J.J. Wu, M. Da Hsieh, W.P. Liao, W.T. Wu, J.S. Chen, Fast-switching photovoltachromic cells with tunable transmittance. ACS Nano (2009). https://doi.org/10.1021/nn900428s
100. J. Lienhard, S. Schleicher, S. Poppinga, T. Masselter, M. Milwich, T. Speck, J. Knippers, Flectofin: a hingeless flapping mechanism inspired by nature. Bioinspir. Biomim. (2011). https://doi.org/10.1088/1748-3182/6/4/045001
101. A. Fazel, A. Izadi, M. Azizi, Low-cost solar thermal based adaptive window: combination of energy-saving and self-adjustment in buildings. Sol. Energy (2016). https://doi.org/10.1016/j.solener.2016.04.008
102. M. Zawidzki, Dynamic shading of a building envelope based on rotating polarized film system controlled by one-dimensional cellular automata in regular tessellations (triangular, square and hexagonal). Adv. Eng. Inform. **29**, 87–100 (2015). https://doi.org/10.1016/J.AEI.2014.09.008
103. C. Doumpioti, E.L. Greenberg, K. Karatzas, Embedded intelligence: material responsiveness in façade systems, in *Life In:Formation: On Responsive Information and Variations in Architecture—Proceedings of the 30th Annual Conference of the Association for Computer Aided Design in Architecture, ACADIA 2010*, pp. 258–262 (2010)
104. S. Alberto, V. Lorenzo, M. Andrea Giovanni, P. Tiziana, An innovative sun shading system for facade application that involves smart materials and novel geometry morphing, in *Facade Tectonics 2020 World Congress* (2020). https://www.facadetectonics.org/papers/shape-memory-alloy-activated-shading!
105. C.K. Khoo, F. Salim, J. Burry, Designing architectural morphing skins with elastic modular systems. Int. J. Archit. Comput. (2011). https://doi.org/10.1260/1478-0771.9.4.397
106. P.R. Denz, C. Sauer, E.F. Waldhör, M. Schneider, P. Vongsingha, Smart textile sun-shading. J. Facade Des. Eng. **9**, 101–116 (2021). https://doi.org/10.7480/JFDE.2021.1.5539
107. Z. Wei, W. Yiqun, S. Menghui, Modeling and simulation of electric-hydraulic control system for bending roll system, in *2008 IEEE International Conference on Robotics, Automation and Mechatronics, RAM 2008*, pp. 661–664 (2008). https://doi.org/10.1109/RAMECH.2008.4681470
108. F. Fiorito, M. Sauchelli, D. Arroyo, M. Pesenti, M. Imperadori, G. Masera, G. Ranzi, Shape morphing solar shadings: a review. Renew. Sustain. Energy Rev. **55**, 863–884 (2016). https://doi.org/10.1016/j.rser.2015.10.086
109. S. Attia, Evaluation of adaptive facades: the case study of Al Bahr Towers in the UAE 8 (2016). https://doi.org/10.5339/QPROC.2016.QGBC.8
110. A. Andy, B. Giorgio, E. David, J. Roy, L. Leonora, L. John, X. Konrad, The Al Bahar towers: multidisciplinary design for Middle East high-rise. Arup J. 90–95 (2013)
111. E. Babilio, R. Miranda, F. Fraternali, On the kinematics and actuation of dynamic sunscreens with tensegrity architecture. Front Mater. **6**, 7 (2019). https://doi.org/10.3389/FMATS.2019.00007/BIBTEX
112. P.T. Martone, M. Boiler, I. Burgert, J. Dumais, J. Edwards, K. MacH, N. Rowe, M. Rueggeberg, R. Seidel, T. Speck, Mechanics without muscle: biomechanical inspiration from the plant world. Integr. Comp. Biol. **50**, 888–907 (2010). https://doi.org/10.1093/icb/icq122
113. S.M. Al-Masrani, K.M. Al-Obaidi, Dynamic shading systems: a review of design parameters, platforms and evaluation strategies. Autom. Constr. **102**, 195–216 (2019). https://doi.org/10.1016/j.autcon.2019.01.014

114. F. Favoino, Q. Jin, M. Overend, Design and control optimisation of adaptive insulation systems for office buildings. Part 1: Adaptive technologies and simulation framework. Energy **127**, 301–309 (2017). https://doi.org/10.1016/J.ENERGY.2017.03.083
115. T. Xenophou, System of using vacuum for controlling heat transfer inbuilding structures, motor vehicles and the like, US patent No. 3,968,831, 1976
116. D.K. Benson, T.F. Potter, C.E. Tracy, *Design of a Variable-Conductance Vacuum Insulation* (1994)
117. A. Berge, C.E. Hagentoft, P. Wahlgren, B. Adl-Zarrabi, Effect from a variable U-value in adaptive building components with controlled internal air pressure. Energy Proc. **78**, 376–381 (2015). https://doi.org/10.1016/J.EGYPRO.2015.11.677
118. T. Pflug, B. Bueno, M. Siroux, T.E. Kuhn, Potential analysis of a new removable insulation system. Energy Build. **154**, 391–403 (2017). https://doi.org/10.1016/j.enbuild.2017.08.033
119. T. Pflug, N. Nestle, T.E. Kuhn, M. Siroux, C. Maurer, Modeling of facade elements with switchable U-value. Energy Build. **164**, 1–13 (2018). https://doi.org/10.1016/j.enbuild.2017.12.044
120. T. Pflug, T.E. Kuhn, R. Nörenberg, A. Glück, N. Nestle, C. Maurer, Closed translucent façade elements with switchable U-value—a novel option for energy management via the facade. Energy Build. **86**, 66–73 (2015). https://doi.org/10.1016/j.enbuild.2014.09.082
121. T. Pflug, N. Nestle, T.E. Kuhn, M. Siroux, C. Maurer, Modeling of facade elements with switchable U-value. Energy Build. (2018). https://doi.org/10.1016/j.enbuild.2017.12.044
122. S. Varga, A.C. Oliveira, C.F. Afonso, Characterisation of thermal diode panels for use in the cooling season in buildings. Energy Build. **34**, 227–235 (2002). https://doi.org/10.1016/S0378-7788(01)00090-1
123. Z. Wu, Z. Feng, B. Sunden, L. Wadsoe, A comparative study on thermal conductivity and rheology properties of alumina and multi-walled carbon nanotube nanofluids. Front. Heat Mass Transf. **5** (2014). https://doi.org/10.5098/HMT.5.18
124. M. Fawaier, B. Bokor, Dynamic insulation systems of building envelopes: a review. Energy Build. **270**, 112268 (2022). https://doi.org/10.1016/J.ENBUILD.2022.112268

Chapter 4
Modelling Tools for Responsive Envelopes

Abstract The aim of this chapter is to provide the reader with the required knowledge to model the responsiveness of innovative envelope technologies. Firstly, the modelling techniques are grouped into two main classes—standard approaches and advanced computational approaches—depending on their availability and complexity. Considering EnergyPlus as reference energy modelling software, the modelling of each responsive technology available is clearly outlined in a step-by-step description to allow the reader to easily follow the modelling activity. With reference to the advanced modelling techniques, both EMS-based approaches and external computational routines are described to provide the widest range of possible applications.

Keywords Energy analysis · Responsive envelope · Façade · Architectural engineering · Building performance simulations

Considering the complexity of the physical phenomena when time-varying properties are adopted, it is extremely difficult to predict the behaviour of responsive devices without requiring computational analyses. Indeed, compared with static envelopes, the responsive devices—depending on their functioning—are much more dependent on the latitude, longitude, weather, users, control strategy, activation thresholds, etc. Furthermore, considering the dynamism of outdoor environment, indoor environment, and envelope properties, the degrees of freedom of responsive models are higher than those considered in traditional buildings. Additionally, these three elements are usually interrelated—depending on the activation criteria—as the envelope properties can change according to indoor and/or outdoor parameters; hence, these dependencies further increase the intricacy of the estimation of the physical parameters involved. Considering, for example, an electrochromic window controlled by solar radiation and indoor temperature sensors, this device depends on both external environmental factors (incident solar radiation) and indoor parameters (indoor air temperature). It follows that both outdoor and indoor environments affect the state of the windows which can, in turn, affect the indoor air temperature and—consequently—the state of the window itself (Fig. 4.1).

F. Carlucci et al., *Responsive Envelopes and Climate Change*, Digital Innovations in Architecture, Engineering and Construction,
https://doi.org/10.1007/978-3-031-58101-4_4

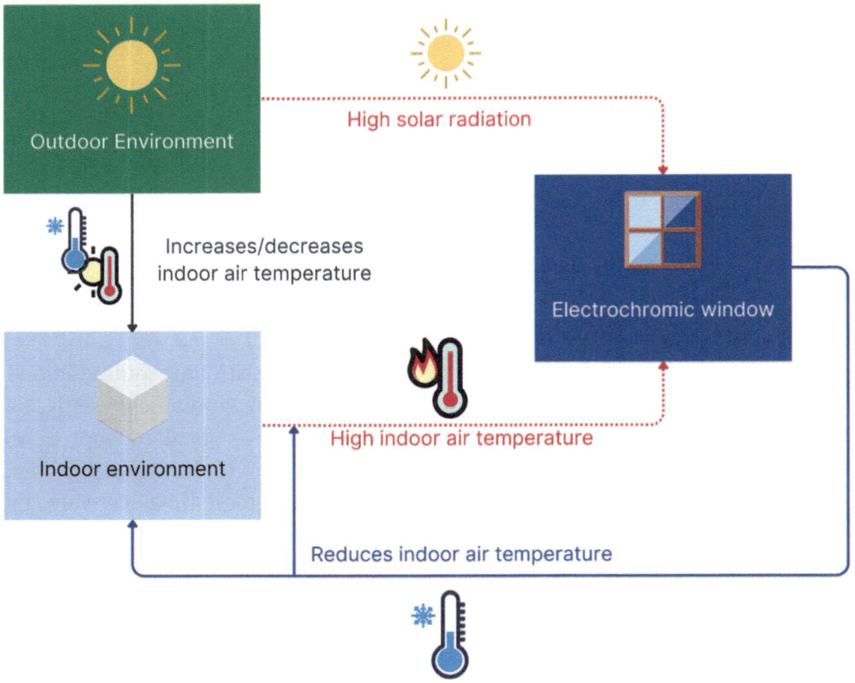

Fig. 4.1 Example of the dependencies in a building energy model equipped with electrochromic windows controlled by solar radiation and indoor air temperature

This scheme briefly explains the need to analyse in detail—through proper energy simulations—all the interactions between the different elements involved to reliably predict the behaviour of the designed building. It is worth noting that, currently, not every responsive system can be easily simulated due to the intrinsic complexity of the phenomena previously described. This chapter aims to provide all the required information to properly model the most diffuse responsive technologies focusing the attention on the different available modelling techniques.

Recently, the rising importance of building energy efficiency has led to the spread of a large number of simulation software to model energy phenomena in the building field; among these the most used are IDA ICE, TRNSYS, EnergyPlus, IES-VE, and ESP-r [1]. ESP-r is an open-source simulation tool based on a finite volume approach; it allows to describe the building from an energy, heat, and moisture point of view. IES-VE is a commercial software largely used in the design field thanks to its easy-to-use interface and to its high integration with sustainability rating systems. TRNSYS is a commercial software based on a graphical environment used to simulate the behaviour of transient systems where users can easily modify existing components or create new ones, extending the capabilities of the environment [2]. IDA Indoor Climate and Energy (ICE) [3] is a commercial software that can model buildings, systems, and its controllers to analyse energy consumption and users' comfort. A deeper insight is provided for EnergyPlus which is the reference tool for this book and is one of the most robust, open source, and widely-used energy analysis tools [4]; moreover, this software includes a set of functions that allows to model different responsive technologies which are the main focus of this work. Another key advantage of EnergyPlus is its multidomain integration and physical interaction; indeed, it considers both thermal, visual, mass-flow, and building service interactions [1, 5] allowing to study the phenomena from different points of view. From a technical perspective, EnergyPlus is a simulation engine created on modular structure codes derived from two software—BLAST (Building Loads Analysis and System Thermodynamics) and DOE–2—respectively developed in 1970s and 1980s [6]. It is not based on a typical user-friendly graphical interface as all the inputs and outputs are comma separated ASCII text files, making complex the modelling of the building. Despite the EnergyPlus idf (input data file) editor allows to easily change an existing model, it is quite difficult to create a new model from scratch. Nevertheless, many dedicated software and plugins allow to easily convert a building model realised through graphical inputs in an EnergyPlus idf file. Design Builder is one of the most diffuse software to create idf file thanks to its user-friendly interface and to its capability of modelling all the building components and systems. It also allows to run EnergyPlus directly from the Design Builder interface—easier than running EnergyPlus standalone—but to access to all the EnergyPlus features needed for responsive technologies, it is often required to export the idf and modify it directly in the idf editor. Alternatively, OpenStudio plugin allows to model the geometry in Sketchup or Rhinoceros and convert them into an EnergyPlus idf. In case of theoretical studies, an alternative starting point for the energy model is proposed by the US Department of Energy (DOE), which developed 16 validated reference building models in different climate zones. These models were based on the Commercial Buildings Energy Consumption Survey (CBECS), ASHRAE (American Society of Heating, Refrigerating and Air-Conditioning Engineers) Standards 90.1, and many other academic sources [7] and can be considered reliable except for few data (lighting power density, electrical equipment, occupancy) which could be considered slightly outdated nowadays in light of technological and functional changes occurred in last years.

Starting from these premises, following sections provide a thorough description of the available modelling techniques divided in existing and innovative computational approaches. The former comprises all those technologies that can be easily modelled through the default available EnergyPlus modules and that do not require any specific programming knowledge. On the contrary, the advanced computational approaches are based on additional scripts—managed directly in EnergyPlus or in external programming software as better described below—that increases the customization of the responsive systems and the number of possible technologies and control strategies that can be simulated. To improve the clarity and for a better understanding, all the subsections are organised by technologies, according to the classification proposed in the previous chapter as shown in Table 4.1.

Table 4.1 Classification of the technologies described in this chapter according to their modelling techniques

Application field	Technology	Standard approach	Advanced computational approach
Improvement of the thermal inertia	*Phase change material*		
	PCM embedded panels	X	
Solar radiation control	*Switchable windows*		
	Electrochromic	X	
	Thermochromic	X	
	Photochromic		X
	Dynamic shadings		
	Simple shadings	X	
	Simplified shadings	X	
	Complex shadings		X
Control of the heat flux	*Variable thermal transmittance devices*		
	Variable U walls	X[a]	X
	Variable U windows		X

[a]Only control strategies based on the envelope temperature can be simulated using this simplified technique

4.1 Standard Approaches

The modelling techniques described in this subsection do not require any specific programming knowledge as they are all based on existing EnergyPlus modules. Clearly, in this case, the technologies and controls available are limited to those provided by the software developers. Nevertheless, EnergyPlus offers a fair number of possibilities to easily simulate responsive devices that can cover several needs in both research and design fields.

4.1.1 Phase Change Materials

With reference to Phase Change Materials, it is worth noting that only PCM embedded in opaque construction can be easily modelled in EnergyPlus. Indeed, despite the software allows to model materials with a specific enthalpy-temperature function, these materials cannot be applied to a windows cavity or glazing. A specific approach, based on a dedicated algorithm, could be adopted to model the PCM windows; nevertheless, considering the limited adoption of this technology and the lower reliability of the calculations, this approach is currently excluded from this book.

Firstly, to model opaque PCM, it is necessary to define the envelope constructions and all their relative materials in the proper EnergyPlus sections ("Material" and "Construction") without any specific foresight. Once the static materials and constructions are set, it is possible to define, for one or more materials, a specific enthalpy-temperature function. When PCM materials are considered, this function includes the so-called enthalpy jump which is representative of the latent heat absorption/release as described in the preceding chapter. These functions can be defined in the EnergyPlus dedicated object "Material Property: Phase Change" through the input of a series of temperatures and their corresponding enthalpies. The software allows to specify up to 16 value pairs and it is suggested to use these slots to cover all the temperature range from low values up to 100 °C. It is fundamental highlighting that the temperatures must be inputted in ascending order; moreover, the resulting function cannot be characterised by negative slopes and the minimum slope is set by the material specific heat. The slope of the function represents the sensible heat storage—typical of all the materials—while the enthalpy jump represents the latent heat storage/release related to the melting/solidification phase of the PCMs (Fig. 4.2). It is fundamental to highlight that the inputs of the "Material Property: Phase Change" class are considered only when the Conduction Finite Difference (ConFD) algorithm is selected in the "Heat Balance Algorithm" class. Therefore it is necessary to change the heat balance algorithm from the default Conduction Transfer Function (CTF) to the Conduction Finite Difference algorithm. Despite the computational time significantly rises with this setting, it considers the discretisation of the surfaces according to the thermal diffusivity of the material, allowing to model PCM and variable conductivity materials [8]. Moreover, to obtain reliable results, it is necessary to increase the number of timesteps from 6 (default) to minimum 12 and maximum 60 timestep per hour; 20 timestep per hour is a good trade-off between reliability and computational demand [9] as demonstrated by numerical and measured data [10, 11]. Considering more thorough hysteresis analyses would produce errors of around 1% [12] instead of a just-few-times-higher error—which can be considered acceptable for yearly studies—to the detriment of the overall computational demand. Unlike other technologies considered in the following paragraphs, the PCMs do not require the modelling of a control strategy thanks to their intrinsically controlled nature. Therefore, once all the above-mentioned settings are reported in the idf file, the latent heat storage can be properly simulated in the software simply running the energy analysis.

4.1.2 Electrochromic Windows

A completely different approach should be considered to model the electrochromic windows which are treated in the software as a sort of shading device. Basically, EnergyPlus considers the coloured state of an EC device as a "shading system" for the clear, not triggered, configuration. In the software, the EC windows can be modelled

Field	Units	Obj1
Name		PCM
Temperature Coefficient	W/m-K2	
Temperature 1	C	-30
Enthalpy 1	J/kg	0
Temperature 2	C	-20
Enthalpy 2	J/kg	11320
Temperature 3	C	-10
Enthalpy 3	J/kg	22640
Temperature 4	C	0
Enthalpy 4	J/kg	33960
Temperature 5	C	10
Enthalpy 5	J/kg	45280
Temperature 6	C	20
Enthalpy 6	J/kg	56600
Temperature 7	C	21
Enthalpy 7	J/kg	61732
Temperature 8	C	22
Enthalpy 8	J/kg	66864
Temperature 9	C	23
Enthalpy 9	J/kg	71996
Temperature 10	C	30
Enthalpy 10	J/kg	79920
Temperature 11	C	40
Enthalpy 11	J/kg	91240
Temperature 12	C	50
Enthalpy 12	J/kg	102560
Temperature 13	C	60
Enthalpy 13	J/kg	113880
Temperature 14	C	70
Enthalpy 14	J/kg	125200
Temperature 15	C	80
Enthalpy 15	J/kg	136520
Temperature 16	C	100

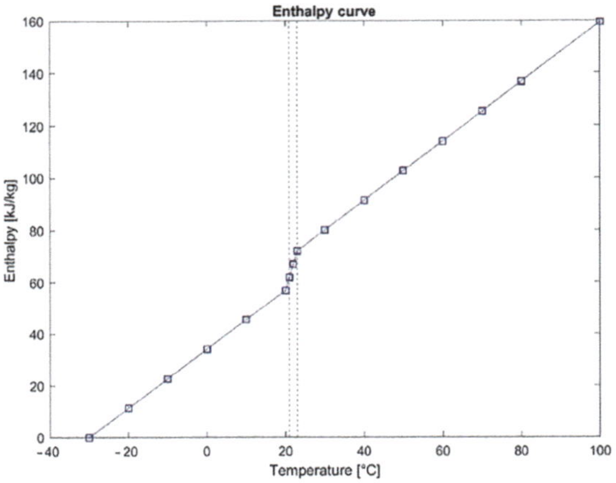

Fig. 4.2 Example of the PCM inputs in the EnergyPlus "MaterialProperty:PhaseChange" module (left) and the corresponding Enthalpy/Temperature function (right)

as switchable windows providing both the clear and the dark state—modelled as separate traditional constructions—and activating a specific window shading control that switches the properties of the windows according to a specific control strategy. The shading effect of the EC film can be simulated using the "Window Shading Control" module that allows to select different shading types. Hence, selecting "switchable glazing" in the dropdown list as shading type, the software switches between the two provided constructions according to the activation criterion selected.

For example, let assume to define two different constructions "Clear Windows" and "Dark Windows" as traditional static windows corresponding to the thermal and optical properties of two states of the EC devices. The unswitched construction (Clear Windows) is assigned directly to the windows surfaces in the "Fenestration Surface: Detailed" class in the "Construction Name" field; on the contrary, the activated state (Dark Windows) is assigned as shading in the "Shading Type" field of the "Window Shading Control" class as shown in Fig. 4.3.

According to the extrinsically controlled nature of the EC devices, in the same module the user can define the control strategy in the "Shading Control Type" field and one or more different activation setpoints. Many predefined control strategies are available in EnergyPlus for the shading—or switchable glazing in this specific case—control; nevertheless, considering that they are mainly referred to the shading systems, additional information is provided in the simple dynamic shading section to avoid redundancies.

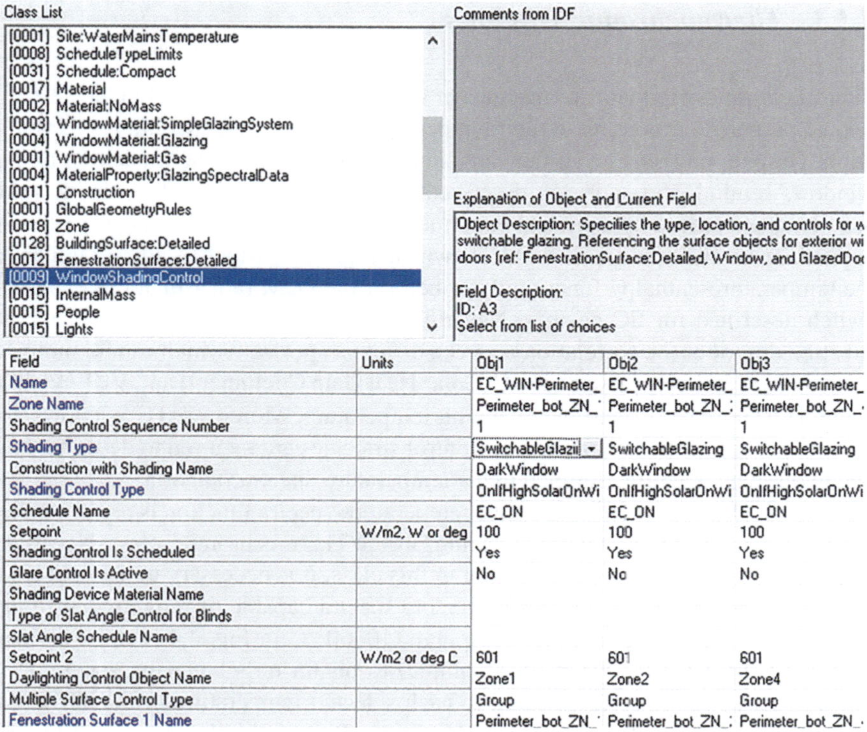

Fig. 4.3 Example of the EC inputs in the EnergyPlus "Window Shading Control" module

According to the provided description, the default EnergyPlus switchable glazing can describe only a two states (on–off) EC windows. Despite in many applications this model can be considered consistent with reality, devices with gradual variation of the thermal properties [13] cannot be modelled. The only exception regards the "Meet Daylight Illuminance Setpoint" control strategy; indeed, when this control strategy is selected, the transmittance of the glazing gradually changes to meet the daylight illuminance setpoint. Nevertheless, using the illuminance control, the device could also assume intermediate states which are not actually available for the considered device. Another important additional control that can be implemented in extrinsically controlled systems regards the EC activation schedule. Indeed, in many cases, it could be useful to limit the functioning of the EC windows to a certain period of the year (e.g., the cooling period) to improve the performance of the building. To consider such control, it is required to create a dedicated programme in the schedule section and select it in the "Schedule name" field of the "Window Shading Control" module.

4.1.3 Thermochromic Windows

Thermochromic windows are responsive systems that can reversibly change their optical properties according to their temperature to control the incoming solar radiation. Despite, on some level, they are similar to EC devices, the modelling of TC windows reminds the approach described for the PCM. Indeed, TC windows are usually characterised by a gradual continuous variation of their optical properties—depending on their temperature as shown in Fig. 4.4—which is more similar to the temperature-enthalpy function described for the PCM rather than to the on–off switch described for EC devices. Nevertheless, unlike the PCMs, in this case the function describes the correlation between different spectra—which can be summed up in the two main properties SHGC (Solar Heat Gain Coefficient) and VLT (Visible Light Transmittance)—and the changing temperature. Moreover, TC windows are intrinsically controlled, therefore, no control strategies are required to describe their behaviour which is fully outlined in the temperature-spectra function.

In EnergyPlus, the definition of the temperature-spectra function is performed by the dedicated "Window Material: Glazing Group: Thermochromic" class. Nevertheless, before modelling the TC function in this class, it is necessary to discretize the TC behaviour and set up a series of glazing spectra capable of reliably describing the function. Hence, the temperature range (10–60 °C in Fig. 4.4) can be divided in n steps (for example 10 in this example) to obtain n + 1 glazing spectra (S1, S2, S3, S4, …, S11) corresponding to each selected temperature of the TC glass. Each spectrum should then be modelled in EnergyPlus as specific glazing ("Window Material: Glazing" class) before modelling the temperature-spectra function. Once all the spectra are modelled, they can be grouped in a single material in the "Window Material: Glazing Group: Thermochromic" module where the user can input a series of temperatures and their corresponding glazing materials to describe the responsive behaviour of the device as shown in Fig. 4.5. In these fields, the software allows to specify up to 16 value pairs and it is suggested to use these slots to cover all the

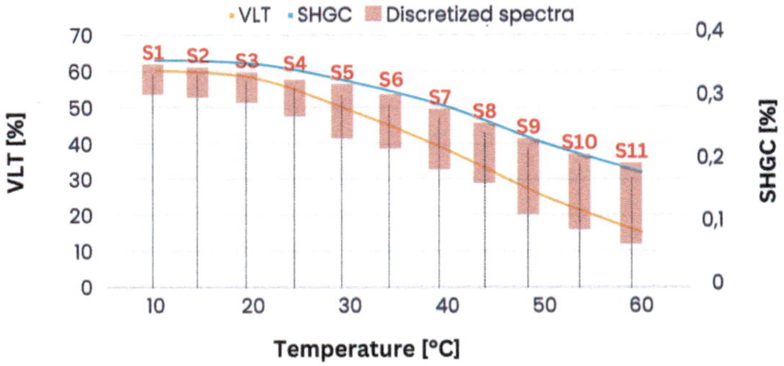

Fig. 4.4 Example of the gradual variation of the TC optical properties related to the glass temperature and possible discretisation of the responsive behaviour in 11 spectra (S1, S2, …, Sn)

temperature range from low values up to 100 °C to create a TC glazing group of spectra.

Finally, the TC windows can be modelled directly in the construction section (similarly to traditional static windows) simply adopting the TC glazing group (namely "TC_Glazing" in Fig. 4.5) as one of the components of the construction. Once the modelling of the function is completed, the responsive behaviour is directly computed by the software during the simulation as, depending on the glazing temperature, it considers the spectral properties associated to that specific temperature during the heat balance calculations.

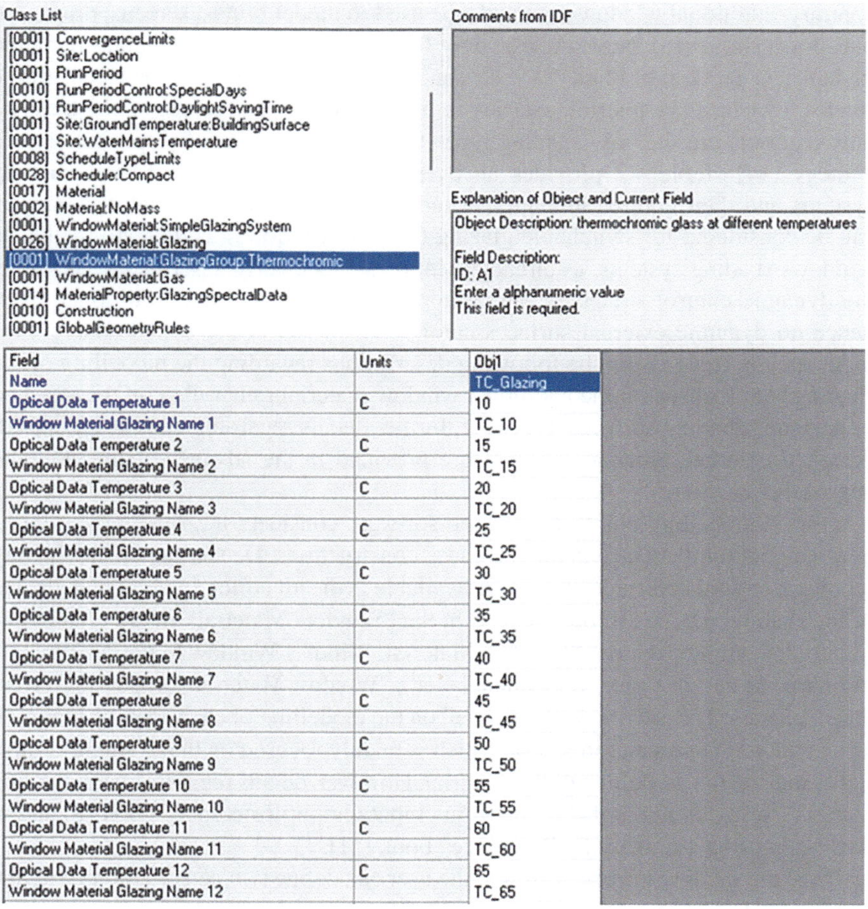

Fig. 4.5 Example of the "Window Material: Glazing Group: Thermochromic" module where the temperature-spectra function is defined. The glazing names TC_10, TC_15 etc. are linked to the "Window Material: Glazing" objects (S1, S2, ..., S11 in Fig. 4.4)

4.1.4 Simple Dynamic Shadings

Regarding solar shading systems, EnergyPlus offers different possible choices to model and control these systems even though with certain limitations. As a general remark, it is important to identify the modelling approach adopted in EnergyPlus for the shading elements before studying their dynamic control. The software classifies a shading element in two main categories: external obstacle to the solar radiation (called attached/detached shading surfaces) or additional window layers (i.e., window material). Trees, other buildings, overhang, fins, etc. can be included in the first class as they can be considered attached or detached shading surfaces; on the contrary, additional window layers can be used to model blinds, screens, drapes, or pull-down shades [8]. Nevertheless, only few of these can be modelled as dynamic systems; in particular, blinds (which can be interior, exterior, or between glass), shades (which can be interior, exterior, or between glass), and screens (which can be only exterior) are the only shading types that can be managed by specific control strategy [14]. Table 4.2 provides an overview of all the possible solar shading systems and their control strategies, where available. The same control strategies can be considered for switchable glazing (i.e., EC windows) as they are treated as windows shading systems, as already mentioned. As clearly described in the table, the dynamic control strategies can be applied only to windows' additional layer, hence no dynamic external surfaces can be considered in the default EnergyPlus modules. For this reason, the following descriptions regarding the modelling of the shadings are focused on the additional window material rather than on the attached or detached shading surfaces. However, further details regarding complex attached/detached external dynamic shadings are provided in the advanced computational approaches section.

From a modelling point of view, the software considers the shading system as new material implemented in the windows' constructions. Depending on the shading typology, several dedicated classes are available in the idf editor. In particular, details of the shading device can be provided in the "Window Material: Shade", "Window Material: Complex Shade", "Window Material: Blind", "Window Material: Screen", "Window Material: Shade: Equivalent layer", "Window Material: Drape: Equivalent layer" classes. This section will not dwell on the modelling of each shading material, as it is out of the scope of this book which is mainly focused on the dynamic control rather than on the modelling of the material. However, details regarding each shading material can be found in the "EnergyPlus Input Output Reference" book [8] and in the "EnergyPlus Engineering Reference" book [11].

Once the shading materials are set, the user can define two constructions with the same glazed components and different shadings. The first one, without any shading material, corresponding to the retracted shading system and, the second one, with the selected shading to represent the activated shading. The retracted state must be directly applied to all the windows surfaces in the "Fenestration Surface: Detailed" (Fig. 4.6) similarly to what already described for the clear state of the EC windows.

Table 4.2 Standard shading devices devices in EnergyPlus v.9.6 [14][1]

Class	Shading typologies	Available control strategies
Attached/Detached shading surfaces	Shading: Site Shading: Building Shading: Site: Detailed Shading: Overhang Shading: Overhang: Projection Shading: Fin Shading: Fin: Projection Shading: Zone: Detailed	• None (Fixed shadings)
Additional window materials	Blinds (interior, exterior, or between glass) Shades (interior, exterior, or between glass) Screens (exterior)	• On if schedule allows • On if high solar on window • On if high horizontal solar • On if high outdoor air temperature • On if high zone temperature • On if high zone cooling • On if high glare • Meet daylight illuminance setpoint • On night if low outdoor temperature and off day • On night if low indoor temperature and off day • On night if heating and off day • On night if low outdoor temperature and on day if cooling • On night if heating and on day if cooling • Off night and on day if cooling and high solar on window • On night and on day if cooling and high solar on window • On if high outdoor air temperature and high solar on window • On if high outdoor air temperature and high horizontal solar • On if high zone air temperature and high solar on window • On if high zone air temperature and high horizontal solar

Hence, the user can select from the "Window Shading Control" module the construction with the shading material in the "Construction with the Shading Name" field, similarly to what happens for the dark state construction in the EC modelling. This description underlines an important limitation of the default module regarding the possible states of the shading systems which can switch their configurations only between two states: retracted or activated. Therefore, no intermediate states can be easily modelled; for example, a partly shaded window can be simulated only splitting the window in two or more parts. Nevertheless, with regards to dynamic systems, splitting the windows in parts could be more difficult to adopt as different part of the same window would be controlled separately and for a proper and advanced control

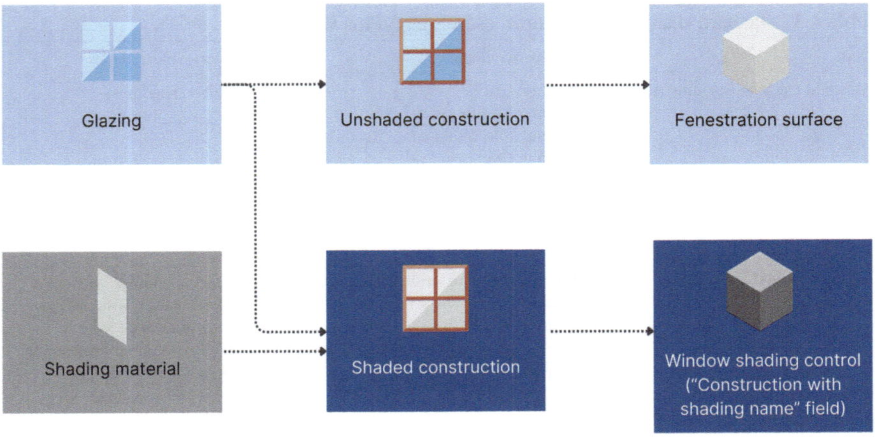

Fig. 4.6 Structure of the modelling of simple dynamic shading through the "Window shading control" module

the users should probably consider additional programming scripts described in the following sections.

Regarding the control strategies, many different approaches—already reported in Table 4.2—are provided by default in EnergyPlus ranging from schedule to radiation control, temperature control, illuminance control, etc. Depending on the strategy adopted, when an activation threshold is required, the users can input in the "Setpoint" field of the "Window Shading Control" module the desired threshold. Few strategies also allow to consider more than a single activation parameter (e.g., "On night if low outdoor temperature and off day", "On if high zone air temperature and high horizontal solar", etc.) creating a combined criterion that allows to activate the shading systems only when both requirements are satisfied. In this case, when two activation thresholds are required, the second one can be inputted in the "Setpoint 2" field. Further details regarding the values of the "Setpoint 1" and "Setpoint 2" fields are provided directly in the idf editor which specifies which parameter is considered in each field. Also in this case, an additional control that could be implemented to manage the dynamic blinds regards the availability schedule, namely, a dedicated schedule that restricts the use of the dynamic shading to a limited period of time (e.g., the cooling period) to improve the performance of the building. To add this control, it is required to create a dedicated program in the schedule section and select it in the "Schedule name" field of the "Window Shading Control" module.

4.1.5 Simplified Dynamic Shadings

The previous section summarises the systems already available in EnergyPlus which are basically related to traditional shading devices. Nevertheless, the significant spread of complex dynamic shadings registered in recent years can hardly be described through these devices. Before moving to advanced detailed computational techniques, a simplified approach is presented as an easier alternative to describe the dynamic devices. This approach is in between the standard existing techniques and the advanced methodologies as it requires different analyses, rather than a single simulation, but it is not strictly necessary to create specific additional computational algorithms to achieve the goal. Despite this method is not the most accurate and shows few discrepancies from the expected behaviour, it allows to easily predict the behaviour of complex systems without specific programming workarounds. Pesenti et al. [15] proposed a methodology to account for the shape variation of the device simply changing the transparency of an equivalent fictitious external screen. With reference to this technique, firstly, it is necessary to analyse the effect of the shading system in different configurations of the device (e.g., fully open, fully closed, 50% folded) in terms of average solar transmittance. Hence the shading effect of the device is summarised—for each shading state—in a series of 8760 hourly solar transmittances. Subsequently, the dynamism of the façade is simulated simply combining these series of hourly values according to specific environmental input (solar height, solar radiation, illuminance, etc.) to consider the change of the shading configuration. Then, these combined transmittance values can be applied as a transparency schedule to a fictitious semi-transparent external vertical surface. Therefore, the complexity of the geometrical definition of the shading system is simplified through the substitution of the device with an equivalent external screen with variable transparency, while the shading responsiveness is settled with a transparency schedule properly defined.

From a practical point of view, this methodology requires a two-steps method summarised in Fig. 4.7. In the first one, a solar radiation analysis is conducted to obtain the values of the solar transmittance for each shading state. This preliminary analysis can be conducted with different software ranging from EnergyPlus to Daysim or Ladybug tools to define the percentage of solar radiation blocked by each shading configuration on an hourly basis. Once all the shading configurations are analysed, these percentage values are properly combined to simulate the façade responsiveness according to an activation criterion, as already described. It is worth noting that in this case the activation criterion cannot be related to any simulation output (e.g., cooling power, indoor air temperature, etc.) because it is estimated before the simulation run; on the contrary, it can be related only to parameters independent of the simulation such as weather data, schedules, etc. Therefore, in this case, the hourly state of the dynamic shading is set in advance and assigned as an input of the simulation. Then, in the second step, these combined values are inputted in EnergyPlus as transparency schedule of the continuous fictitious vertical external shading. In this case, no specific dynamic controls are required in EnergyPlus as they are already embedded in the values of the transparency schedule. Therefore, the continuous vertical shading system can be modelled simply as an offset of the window in the "Shading Building: Detailed" module considering the combined transparency schedule in the "Transmittance Schedule Name" field. The primary limitation of this methodology is related to a numerical discrepancy between real and fictitious shading

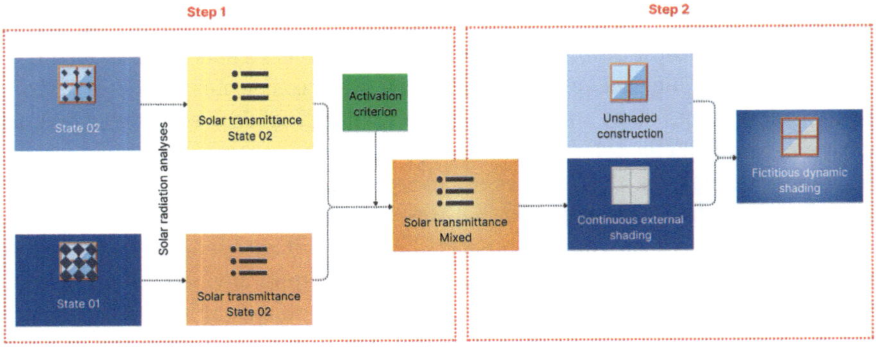

Fig. 4.7 Structure of the modelling of simplified dynamic shading

systems. Indeed, detailed analyses found that the fictitious continuous screen interfere with direct and diffuse radiation, partly obstructing the solar radiation, even when completely transparent [14]. Moreover, the use of activation criteria independent of the simulation outputs clearly reduces the application field of this this modelling technique.

4.1.6 Variable Thermal Transmittance Walls

Envelope solutions with variable thermal transmittance can be modelled both using existing EnergyPlus module or using additional customised scripts. In this specific case, the choice strictly depends on the technology itself as the existing module allows to model only a limited number of devices controlled by the envelope temperature. In particular, the responsive envelope can be assimilated to a standard multilayer construction where one layer is characterised by a variable thermal conductivity. Similar to PCMs, the variable thermal conductivity is expressed as an additional property of a specific material. Hence, after modelling the construction using standard static materials, the variation of the thermal properties can be described in the "Material Property: Variable Thermal Conductivity" module where the users can define a temperature-conductivity function using up to ten value pairs. Also in this case, the inputs of the "Material Property: Variable Thermal Conductivity" class are considered only when the Conduction Finite Difference algorithm is selected in the "Heat Balance Algorithm" module. Therefore it is fundamental to change the calculation algorithm from the default CTF to the ConFD algorithm to properly consider the responsive behaviour during the simulation. Moreover, to ensure a good reliability of the results, it is suggested to increase the number of timesteps from 6 (default) to minimum 12 and maximum 60 timestep per hour. Despite this modelling technique is extremely easy and fast, this approach is clearly very limiting due to its univocal temperature-based activation criterion. Therefore, to consider different simple criteria—e.g., schedule, heating period, etc.—or more complex control strategies, additional customised scripts should be implemented as better explained in the next section. Moreover, this approach considers only the thermal transmittance variation of the wall, which can be considered reliable, for example, for movable layers devices. Instead, further effects related to specific technologies—such as for example the ventilation in permeodynamic systems—cannot be directly implemented with this approach; therefore, they should be properly accounted and modelled as additional components.

4.2 Advanced Computational Approaches

This section provides modelling hints for all those technologies that require additional computational algorithms to properly simulate the energy behaviour of the building. These approaches require different levels of programming knowledge depending on the solutions considered. Two main subclasses of modelling techniques can be identified: the first one—the easiest—is based on the internal EnergyPlus programming module called Energy Management System (EMS), while the second one is based on external programming software to create additional routines to allow complex simulations. Technologies such as photochromic, variable U windows can be modelled through the existing EnergyPlus EMS while complex dynamic shadings require additional external computational routines to be properly modelled. Clearly, in these classes we could also include all the technologies covered by existing EnergyPlus modules—described in the previous sections—which could also be modelled with additional computational routines to add further details or specific control strategies. Therefore, this section could, potentially, includes all the possible responsive envelope technologies. Considering the wide scope and the complexity of the programming language, next paragraphs aim to provide the general approach and layout of modelling techniques without describing in depth the programming syntaxes which are considered out of the scope of this book.

4.2.1 EMS-based Modelling Techniques

Considering the first approach, the users can freely include additional algorithms in the EnergyPlus simulation, writing in the EMS module proper strings of code. This approach is based mainly on three components: the sensors, the code itself, and the actuators. One of the key points of these techniques is the use of a wide range of virtual sensors to monitor one or more parameters during the simulation in order to define specific customisable actions [16]. The EMS sensors transform the variables and parameters used during the EnergyPlus simulation (e.g., the indoor temperature, occupancy status, solar radiation, etc.) in inputs for an additional customised code created in the EMS. Hence, starting from these EMS inputs, the user can define a certain number of operations to describe—in the responsive envelope field—the control strategy adopted. Once the elaboration of the inputs is completed, the EMS can change schedules, envelope properties, etc. using the EMS actuators that can vary the starting EnergyPlus simulation inputs. The EMS is, hierarchically, a high-level control method; it can potentially turn on and off pumps, fan, lights, change setpoints, envelope properties, etc. during the simulation according to specific scripts. Nevertheless, in this book all the examples provided are related only to the techniques adopted to model responsive building envelope.

From a programming point of view, the EMS adopts the EnergyPlus Runtime Language (ERL) to describe the control routines. Basically, the sensors declare the ERL variables, which can be selected among the EnergyPlus output variables or meters. This definition is set in the "Energy Management System: Sensor" class where the user can define the name of the sensor ("Name" field), the output variable or meter selected ("Output: Variable or Output: Meter Name" field), and the key reference for the specified output variable to identify the specific output requested ("Output: Variable or Output: Meter Index Key Name" field). For example, to read the temperature of a specific zone, it is necessary to select "Zone Air Temperature" as "Output: Variable or Output: Meter Name" and the name of the selected zone as "Output: Variable or Output: Meter Index Key Name". On the contrary, the key name must be omitted for weather parameters and meters as they do not have keys. The list of the possible outputs and meters can be easily checked in the eplusout.rdd and eplusout.mdd output files (which are automatically generated after a preliminary simulation, regardless of the creation of an EMS program) to select proper sensors. Similarly, the EMS actuators identify an ERL variable to change accordingly the value of an actuated component (schedule value, construction, etc.). The actuators are defined in the "Energy Management System: Actuator" class where the user can define: their name ("Name" field), the name of the actuated component (for example the name of a specific window in the field "Actuated Component Unique Name"), its type (e.g., surface for the window example in the field "Actuated Component Type"), and its control type (e.g., construction state for the window example in the field "Actuated Component Control Type"). A list of the possible actuators can be found in the eplusout.edd file which is created running a simulation after selecting it as output in the "Output: Energy Management System" module. Finally, regarding the script itself, it can be created in the "Energy Management System: Program" class. In this module, the user—after defining the program name in the first cell—can write a series of program lines using the ERL which is a specific complete programming language with its own syntax. To avoid taking too long on the ERL syntax, the readers can refer to the dedicated EnergyPlus guide [16] for a detailed explanation, while in the next subsections specific precompiled scripts for responsive envelopes will be given as examples. Once the sensors, the programs, and the actuators are defined, the programs should be called in the "Energy Management System: Program Calling Manager" to specify when each EMS program is run.

4.2.1.1 Photochromic Glazing

Despite the effects of the PC windows can be considered similar to solar-activated EC devices, the default EnergyPlus switchable glazing module—usually adopted for EC devices—cannot be considered suitable for PC due to its on/off behaviour. Indeed, the PC films gradually change their optical properties according to the incident solar radiation [17]; therefore, the concept behind this approach is more similar to what already exists in the EnergyPlus default modules for thermochromic materials. Nevertheless, in this case, the responsiveness is expressed as a radiation-spectra function rather than a temperature-spectra dependency.

Similarly to the TC windows, the first step of this approach is the discretisation of the responsive behaviour creating a series of glazing spectra. Then, each spectrum is modelled in EnergyPlus as a single glazing ("Window Material: Glazing" class) and implemented in a specific construction before modelling the radiation-spectra function. For example, let assume to discretise the spectral variation in ten different spectra considering steps of 100 W/m^2; each spectrum corresponds to a specific glazing which is, in turn, adopted as one of the layers of an EnergyPlus construction, obtaining ten different window constructions which represent the different states of the window under different radiation conditions. Once this preliminary standard modelling is completed, it is possible to represent the function in the EMS modules (Fig. 4.8).

Firstly, it is necessary to create an EMS sensor for each window to read the value of the incident radiation. To that end, in the "Energy Management System: Sensor" module the user can create new sensors specifying: the sensor name, the name of the monitored window as key (in the "Output: Variable or Output: Meter Index Key

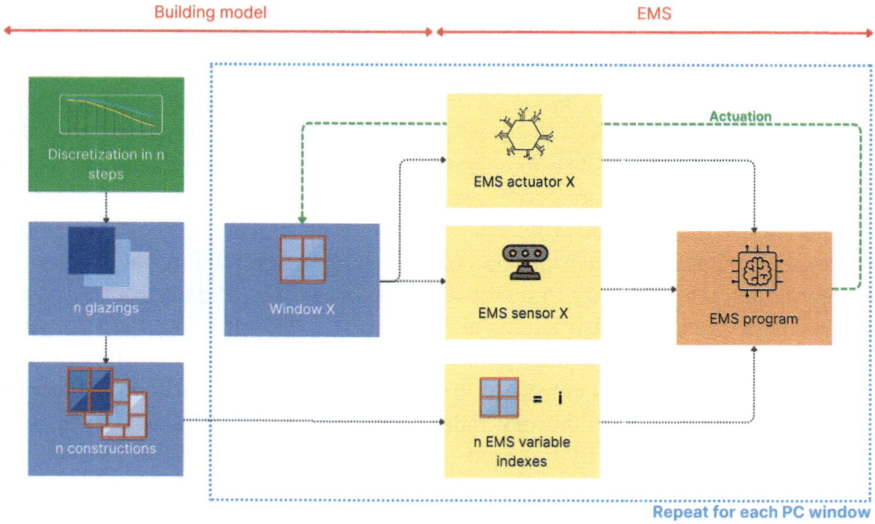

Fig. 4.8 Schematic description of the EMS approach for PC windows

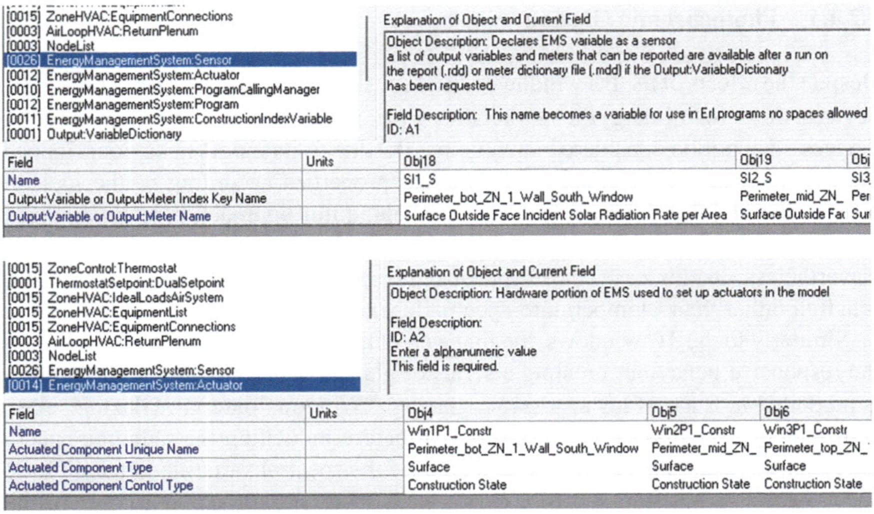

Fig. 4.9 Example of the "Energy Management System: Sensor" (above) and "Energy Management System: Actuator" module (below)

Name" field), and "Surface Outside Face Incident Solar Radiation Rate per Area" in the "Output: Variable or Output: Meter Name" field. Figure 4.9 shows an example of the modelling of EMS sensors for a PC window where SI1_S, SI2_S, etc. are the sensors associated to each PC window of the building, "Perimeter_bot_ZN_1_Wall_South_Window" is the specific name of the monitored window in this example, and the last field reports the EnergyPlus output variable selected.

After defining all the EMS sensors of the building, the EMS actuators—one for each window—can be defined in the "Energy Management System: Actuator" class (Fig. 4.9). Each actuator refers to a window already associated with a sensor in the "Actuated Component Unique Name" (in the example both the sensor "SI1_S" and the actuator "Win1P1_Constr" refer to the window "Perimeter_bot_ZN_1_Wall_South_Window"). The remaining fields "Actuated Component Type" and "Actuated Component Control Type" are always, respectively, "Surface" and "Construction State" for all the PC windows because they depend only on the type/control of the component rather than on the building model itself. Before moving to the EMS program, it is necessary to set the last preliminary class regarding the "Energy Management System: Construction Index Variable"; this module creates an ERL variable containing an index that points to the selected construction. Basically, it transforms the EnergyPlus construction in a suitable index for ERL; therefore, it simply contains the name of this new index (in this example GlzSys_0, GlzSys_100, …, GlzSys_1000) associated to the existing standard EnergyPlus construction. The new index and the original standard construction can also have the same name to ease the reading and interpretation of the file.

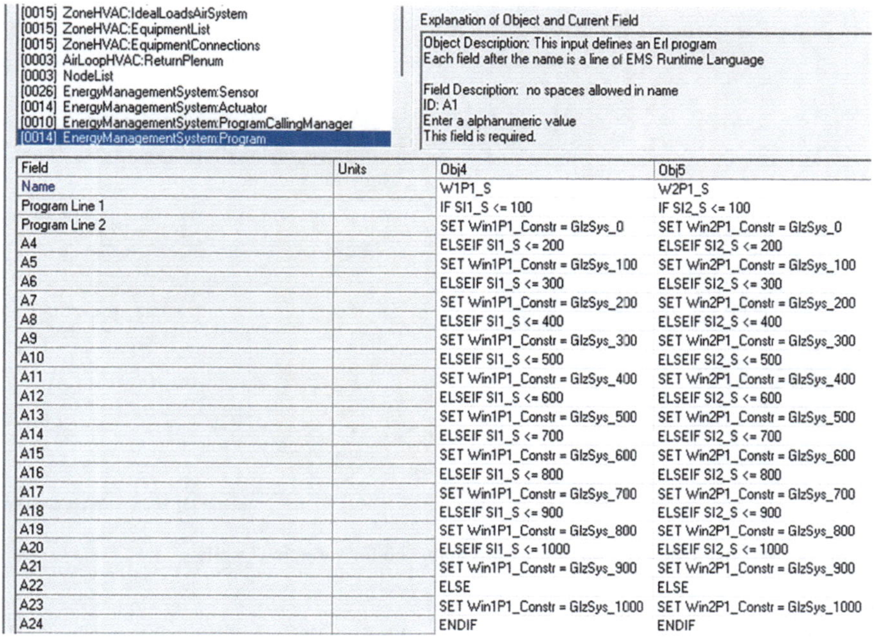

Fig. 4.10 Example of the "Energy Management System: Program" class

These preliminary steps are required to write the EMS program which is described in detail in Fig. 4.10. It is suggested to define a specific program for each window for sake of simplicity; for example, in the following figure (Fig. 4.10) the "W1P1_S" program refers univocally to "Perimeter_bot_ZN_1_Wall_South_Window" construction through the "SI1_S" sensor and the "Win1P1_Constr" actuator. The proposed code simply compares—through a series of if/elseif/else control structures—the incident solar radiation on a window with predefined thresholds (100 W/m^2 steps in this example) derived from the preliminary discretisation of the responsive behaviour. Hence, it associates the corresponding construction to the window through the "set" command applied to the actuator and the construction index variable (basically: "set [actuator name] = [construction index variable name]"). Finally, all the EMS programs are reported in the "Energy Management System: Program Calling Manager" adopting "Begin Timestep Before Predictor" as calling point.

4.2.1.2 Variable Thermal Transmittance Walls

For all the variable U walls which are not triggered by the envelope temperature—already described in the previous section—, a specific EMS program can be developed to simulate their behaviour. To that end, the user can adopt the same approach described for the PC windows simply connecting the sensors and actuators to wall

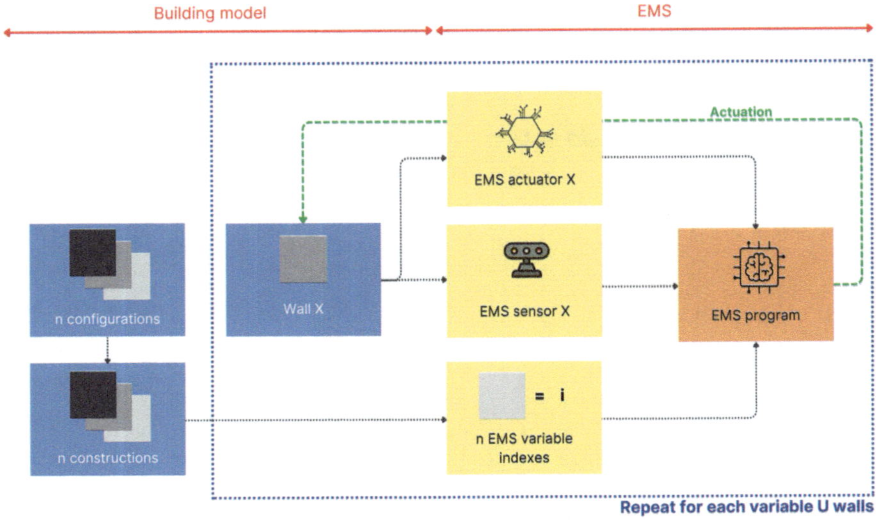

Fig. 4.11 Schematic description of the EMS approach for variable U walls

surfaces rather than to windows surfaces. Unlike PC windows, for extrinsically controlled devices it is not possible to define a general algorithm that describes the technology as it strictly depends on the control strategy adopted. Therefore, in this case, no specific algorithms will be provided but the following general scheme (Fig. 4.11) can give the bases to develop the customised scripts.

Firstly, it is necessary to define one static construction for each state of the responsive wall; then the approach is overlapping to what already described for the PC windows. The only care is to write the wall name in the "Actuated Component Unique Name" of the actuators. Instead, the "Output: Variable or Output: Meter Name" and "Output: Variable or Output: Meter Index Key Name" fields of the sensors depend on the control strategy adopted; for example, if it is activated by the indoor air temperature the variable name is "Zone Air Temperature" and the key is the name of the thermal zone monitored. Further details on possible output variable are provided in the eplusout.edd file as already described at the beginning of this section. It is worth noting that this approach should be carefully validated because changing the construction during the simulation implies differences in the surface temperature which strictly depends on their thermal inertia. Therefore, this approach works better on low thermal inertia constructions because they are less affected by the transient nature of the thermal transfer. However, before running the energy analyses, it is highly suggested to test and validate the script, analysing the variables (surface temperature, air temperature, etc.) involved during the simulation.

4.2.1.3 Variable Thermal Transmittance Windows

Variable thermal transmittance windows can be easily modelled adopting the same approach described for the variable U walls. In this case, the actuators refer to the window name in the "Actuated Component Unique Name". Unlike the opaque walls, in this case, the low inertia of the glazed structures does not imply specific issues in the window surface temperature. Theoretically, these devices could be modelled also as switchable glazing using the same SHGC and VLT and different U values. Nevertheless, previous studies demonstrated that, currently, adopting the approach described for the EC windows for variable U windows leads to unreliable results.

4.2.2 External Computational Routines

This class includes all those approaches that still rely on EnergyPlus as simulation engine but combine this software with external algorithms that allow to run complex routines and manage the energy simulations. The concept behind this approach is similar to the EMS-based technique but it can be considered a more powerful evolution of the ERL language as it can access to a series of libraries, debugging functions, and more complex programming tools. Thanks to the introduction of the EnergyPlus API (Application Programming Interface), EnergyPlus can be included in a series of new workflows and applications increasing its potential. Moreover, thanks to the runtime API, external functions can be called back at certain points of the simulation. It is clear that the potential of this approach is extremely wide and is strictly connected to the specific application and study conducted. In this book, this approach will be considered only to show a simulation approach—proposed and validated in a previous study [14]—to model complex external dynamic shading systems. This modelling technique run a series of simulations and elaborate their results directly from a Python code. In this case, no direct access to the EnergyPlus variables is required during the simulation run; therefore, the modelling is slightly simpler even though it requires different computational tasks. The management of the EnergyPlus idf can be done in different ways; for example, in the above-mentioned study, the Python library "Eppy" was adopted to access and modify the idf. Considering the length and complexity of the algorithm, the following paragraph describes, from a conceptual perspective rather than from a programming point of view, the suggested modelling approach without describing in detail the algorithm itself which can be developed in different ways according to the user programming languages.

4.2.2.1 Complex Dynamic Shading Systems

The main issue of the complex dynamic shading systems is that they cannot be directly modelled in EnergyPlus. Even changing the shading geometry (vertex coordinates) during the simulation through EMS or additional algorithms, the shadowing analysis is run at the beginning of the simulation and, therefore, following changes in the shading shapes are not taken into account and do not affect the simulation. For this reason, traditional approaches usually run different simulations—one for each shading configuration—considering the shading fixed in a specific configuration; then, the output obtained are usually combined according to a specific activation criterion. These modelling techniques can be considered reliable for those physical parameters that change instantaneously—i.e., illuminance, glare, etc.—while thermal and energy parameters need a single continuous simulation to reliably describe their transient nature. For example, considering traditional approaches, when the shading system changes from state 1 (timestep t) to state 2 (timestep t + 1), both illuminance and air temperature of the thermal zone assume the values of the state 1 model at timestep t and state 2 model, at timestep t + 1. While the values of illuminance are reliable as they depend only on the shading geometry, the air temperature is less accurate as it is based on the values of the previous timesteps (t − 1, t − 2, etc.) which can be significantly different between the two models (for example 20 °C for the less shaded model and 16 °C for the shaded model). Traditional approaches combine the output of the two models overlooking that the previous timesteps can influence the current value; the resulting output is characterised by values that unreliably "jump" from state 1 to state 2 model values (in the previous example from 16 to 20 °C in a 5 min timestep) neglecting the transient nature of thermal parameters. Hence, the main innovation of the study proposed [14] is that it can properly accounts for the continuous nature of the thermal and energy phenomena. Indeed, rather than working on combined outputs, this innovative algorithm combines the inputs—in particular the sunlit fractions calculated for each shading state—and forces the simulation with an externally combined input file (external shading file). This approach assures that the energy behaviour of the system is calculated as the output of a single run—rather than as combination of different simulations—where the shape variations of the shading system lead to gradual and continuous variations in the thermal state of the building elements. The core of this approach is the EnergyPlus shading calculation file which is a csv file containing all the sunlit fractions (i.e., the shading ratios) of all the surfaces of the model in all the simulation timestep. This file can be easily generated in EnergyPlus in the "Shading calculation" module activating the field "Output External Shading Calculation Results" and contains all the information regarding the sunlit fraction of each surface during the whole simulation. These files can be generated for each of the shading configuration and then combined—through external algorithms—in a single file that represent the dynamic behaviour of the shading system. Hence, this file can be implemented in an equivalent EnergyPlus model (without any real shading) as an external precomputed schedule using the 'Schedule: File: Shading' class. The above-mentioned study validated this workflow and demonstrated that using this external schedule on an equivalent model provides reliable results with many different complex shading systems.

From a practical point of view, the energy behaviour of the model is described starting from two different model types:

- A Static Shading Model (SSM) for each configuration of the shading system: in these static models the shading systems are fixed and cannot change during the simulation. These models are used to calculate the shading effect—i.e., the shading fractions—of each of the shading configuration on the model surfaces. Subsequently, these shading fractions files are combined, through external algorithms, in a Dynamic Shading File (DSF) according to the activation criterion and used in the equivalent model.
- An Equivalent Dynamic Shading Model (EDSM) for the final simulation (Fig. 4.12): in this file no real shadings are provided as the shading calculation is forced with the external DSF to simulate the change of shape of the shading system. In this model the thermal parameters can vary continuously considering that the dynamic effect of the shading system is already embedded in the DSF.

This approach requires a minimum number of n + 1 total simulations, where n is the number of possible shading configurations. It is reliable from a thermal point of view but, if daylighting or automated lighting controls are considered, the lighting consumption and illuminance data must be deducted from the SSMs because the shading file affects only the thermal analyses. Nevertheless, these data can be easily combined in external algorithms through traditional approaches because, as already explained, they change instantaneously.

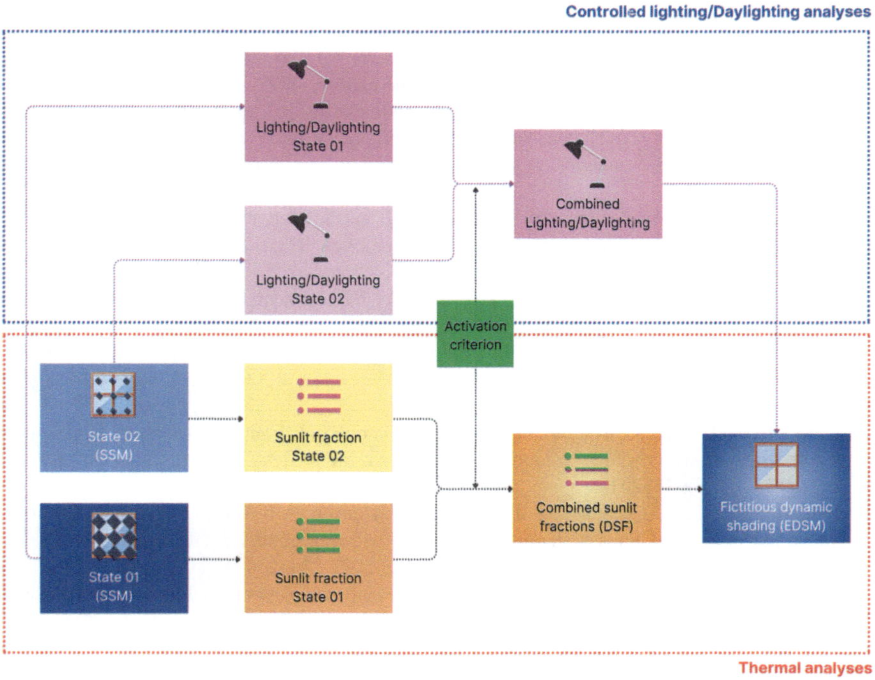

Fig. 4.12 Schematic description of the proposed algorithm for complex dynamic shading systems

The main limitation of this technique is that it works only using activation thresholds independent from thermal parameters of the thermal zones (such as indoor temperature, cooling/heating state, etc.) because the combination of the shading files is set before the final simulation.

References

1. F. Favoino, L. Giovannini, R. Loonen, Smart glazing in Intelligent Buildings: what can we simulate? in *GPD 2017—Glass Performance Days 2017*, pp. 8–15 (2017)
2. Welcome | TRNSYS : Transient System Simulation Tool (n.d.). https://www.trnsys.com/. Accessed 22 Sept 2022
3. IDA ICE—Simulation Software | EQUA (n.d.). https://www.equa.se/en/ida-ice. Accessed 22 Sept 2022
4. V. Corrado, E. Fabrizio, Steady-state and dynamic codes, critical review, advantages and disadvantages, accuracy, and reliability, in *Handbook of Energy Efficiency in Buildings: A Life Cycle Approach*, pp. 263–294 (2019). https://doi.org/10.1016/B978-0-12-812817-6.00011-5
5. R.C.G.M. Loonen, F. Favoino, J.L.M. Hensen, M. Overend, Review of current status, requirements and opportunities for building performance simulation of adaptive facades†. J. Build. Perform. Simul. **10**, 205–223 (2017). https://doi.org/10.1080/19401493.2016.1152303
6. U.S. Department of Energy, EnergyPlus™ Version 22.2.0 Documentation. Getting started (2022)
7. K. Field, M. Deru, D. Studer, Using doe commercial reference buildings for simulation studies, in *Proceedings of the SimBuild 2010 Fourth National Conference of IBPSA-USA* (2010)
8. US Department of Energy, Input Output Reference. EnergyPlus (2021). https://energyplus.net/assets/nrel_custom/pdfs/pdfs_v9.6.0/InputOutputReference.pdf
9. F. Carlucci, A. Cannavale, A.A. Triggiano, A. Squicciarini, F. Fiorito, Phase change material integration in building envelopes in different building types and climates: modeling the benefits of active and passive strategies. Appl. Sci. **11**, 4680 (2021). https://doi.org/10.3390/APP11104680
10. P.C. Tabares-Velasco, C. Christensen, M. Bianchi, Verification and validation of EnergyPlus phase change material model for opaque wall assemblies. Build. Environ. (2012). https://doi.org/10.1016/j.buildenv.2012.02.019
11. U.S. Department of Energy, EnergyPlus Version 22.1.0 Documentation: Engineering Reference (2022)
12. A. Zastawna-Rumin, T. Kisilewicz, U. Berardi, Novel simulation algorithm for modeling the hysteresis of phase change materials. Energies (Basel). **13**, 1200 (2020). https://doi.org/10.3390/en13051200
13. S. Sibilio, A. Rosato, M. Scorpio, G. Iuliano, G. Ciampi, G.P. Vanoli, F. De Rossi, A review of electrochromic windows for residential applications. Int. J. Heat Technol. (2016). https://doi.org/10.18280/ijht.34S241
14. F. Carlucci, R.C.G.M. Loonen, F. Fiorito, J.L.M. Hensen, A novel approach to account for shape-morphing and kinetic shading systems in building energy performance simulations, pp. 1–20 (2022). https://doi.org/10.1080/19401493.2022.2142294
15. M. Pesenti, G. Masera, F. Fiorito, Exploration of adaptive origami shading concepts through integrated dynamic simulations. J. Archit. Eng. (2018). https://doi.org/10.1061/(asce)ae.1943-5568.0000323

16. U.S. Department of Energy, EnergyPlus Version 22.1.0 Documentation: Application Guide for EMS (2022)
17. A. Cannavale, G. Zampini, F. Carlucci, M. Pugliese, F. Martellotta, U. Ayr, V. Maiorano, F. Ortica, F. Fiorito, L. Latterini, Energy and daylighting performance of building integrated spirooxazine photochromic films. Sol. Energy **242**, 424–434 (2022). https://doi.org/10.1016/J.SOLENER.2021.10.058

Chapter 5
Design Strategies

Abstract This chapter considers a reference office building to apply all the theoretical concepts described in the previous sections. In these practical examples, three main responsive technologies—phase change material, electrochromic windows, and variable thermal transmittance windows—are considered to test the benefits of these devices in both current and future climate scenarios. Moreover, to account for their weather dependency, each technology is tested in three locations characterised by extremely different weather (Larnaca, Rome, Helsinki). The obtained outcomes allow to bring out possible weaknesses and strengths of responsive technologies in different scenarios, to properly understand their functioning and maximise their efficiency.

Keywords Responsive envelope · Electrochromic · Phase change material · Climate change · Architectural engineering

Previous chapters provided all the required starting knowledge to understand, classify, and model a responsive technology; to complete this review, this chapter aims to provide hints to properly design these technologies. As already mentioned, it is difficult to predict and generalise the behaviour of these systems due to their high degrees of freedom. Moreover, extrinsically controlled systems can thoroughly change their behaviour simply varying the activation criteria and thresholds adopted. Therefore, this section does not aim to provide general stiff rules to design these systems in every possible environmental and climate context as this approach would be limiting—considering the broad range of possible solutions—and inconsistent with the nature of the technology itself. Instead, the main goal is to test their potential in different sample contexts to bring out a new awareness of eventual weaknesses and strengths that can help designers and researchers to understand their functioning and to maximise the benefits related to these devices.

All the assessments shown in this chapter are conducted on a sample medium office model based on the DOE reference EnergyPlus model (Fig. 5.1) and converted into a set of models, each of which consistent with current energy efficiency laws and

F. Carlucci et al., *Responsive Envelopes and Climate Change*, Digital Innovations in Architecture, Engineering and Construction,
https://doi.org/10.1007/978-3-031-58101-4_5

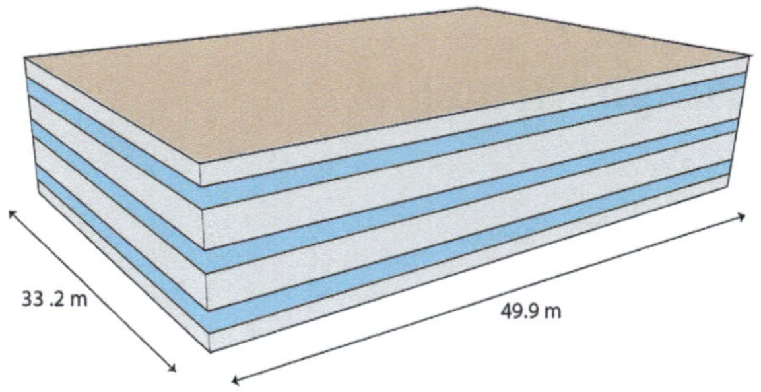

Fig. 5.1 Medium office reference model geometry

with the locations described below. In particular, the selected technologies are tested on all the facades—with the exception of the northern façade for the EC windows—to understand the overall contribution to the energy efficiency of buildings. The choice of the office building type is related to its standardised operational schedules—compared to other types, such as residential buildings—that reduces the number of possible variables of the problem.

Considering that the functioning of responsive technologies strictly depends on the analysed locations, in this section different climates are taken into account to highlight possible critical issues related to specific climate zones. Specifically, three strongly different locations—Larnaca, Rome, and Helsinki, whose climates are described in Fig. 5.2—were identified to conduct these analyses. The three locations selected differs significantly considering both Heating Degree Days (HDD) and Cooling Degree Days (CDD). Hence, each configuration studied was tested in a warm (Larnaca: $CDD_{18^\circ} = 1259$, $HDD_{18^\circ} = 759$), in a cold (Helsinki: $CDD_{18^\circ} = 33$, $HDD_{18^\circ} = 4712$), and in an intermediate climate (Rome: $CDD_{18^\circ} = 649$, $HDD_{18^\circ} = 1444$) with a high variation of the main environmental parameters.

Moreover, different future climate scenarios are generated to understand how these systems can react to long term weather variations. Indeed, although these devices were initially designed to respond to short to medium term changes of the external environment (sub-hourly, hourly, daily, seasonal), an emerging trend is to assess their potential over longer periods (decades) to increase the building resilience to climate change. Therefore, starting from the Typical Meteorological Year (TMY) weather files of the three above mentioned locations, future climate scenarios were generated using Future Weather Generator. In order to test the responsiveness of the selected technologies under a changing environment, the future weather scenarios are referred to the furthest timeframe (2066–2095, called 2080) and to the highest emission scenario (SSP5-8.5) to maximise the differences with the current typical scenario (Fig. 5.3).

In order to avoid the dependency of the consumption results on the HVAC systems, the default auto-sizing HVAC systems were substituted with properly calibrated ideal loads air systems. The ideal loads systems are included among the EnergyPlus systems and occupy the same hierarchical place of the auto-sized HVAC units.

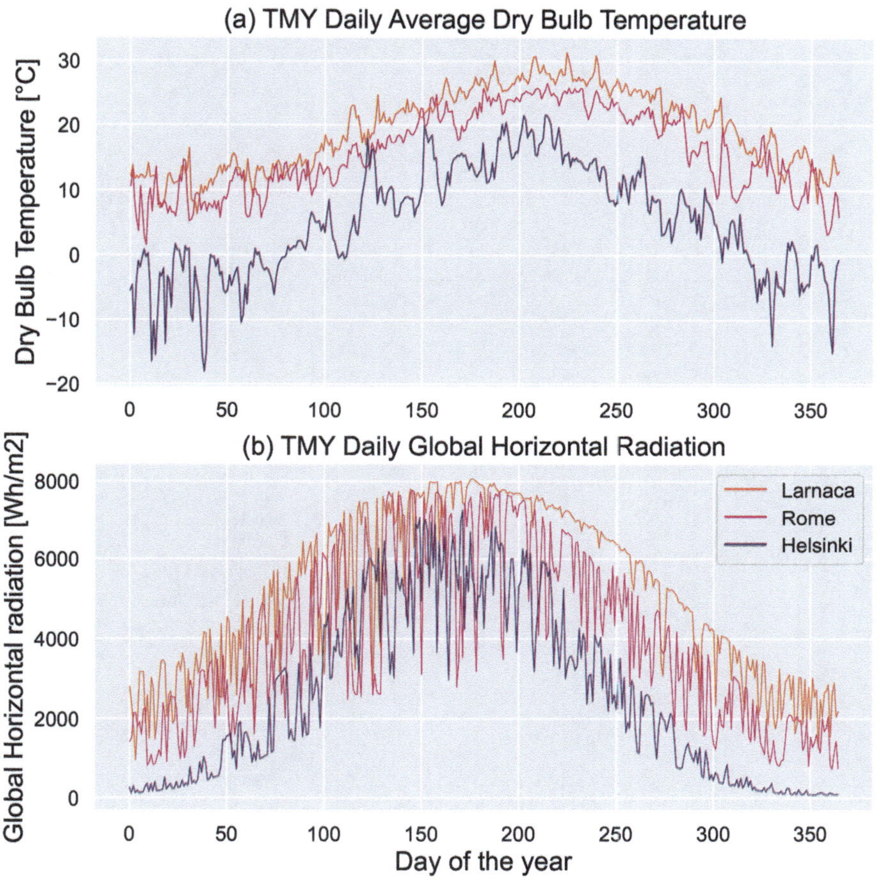

Fig. 5.2 Climatic characteristics of the selected locations: **a**) daily average dry bulb temperatures, **b**) daily cumulative global horizontal radiation. Data based on long term real observation deducted from Energy Plus Weather file

Nevertheless, these systems are not connected to any real generation and distribution network, but each one can directly supplies heating or cooling to satisfy the zones setpoints. Considering that these ideal systems do not include the heating/cooling generation process, they directly provide the thermal demand rather than the consumption. To obtain the energy consumption, in this study an ideal Energy Efficiency Ratio (EER) equal to 3 in cooling mode and a Coefficient of Performance (COP) equal to 3 in heating mode were considered.

Bearing in mind the wide range of responsive technologies described, a representative technology was selected for each functioning class, according to what already described in the previous sections. In particular, a PCM embedded plasterboard, an EC window, and a variable-U window are considered as representative technologies for, respectively, the improvement of thermal inertia, the control of solar radiation,

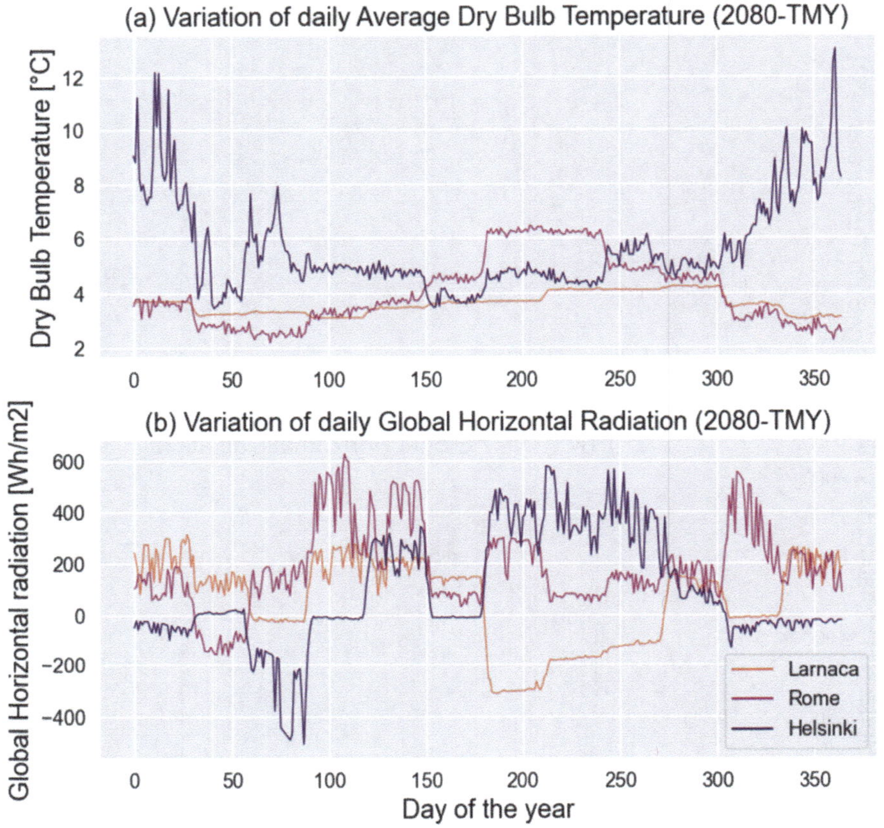

Fig. 5.3 Variations (2080-TMY) of the climatic characteristics of the selected locations: **a**) daily average dry bulb temperatures, **b**) daily cumulative global horizontal radiation

and the control of the heat flux. Table 5.1 sums up the simulation framework considered for this chapter. Starting from these general inputs, specific inputs for each technology were selected to cover different possible cases. Hence, three different activation criteria were considered for the extrinsically controlled technologies (electrochromic and variable-U windows) and two different materials and two different positions in the multilayer wall were adopted for the intrinsically controlled systems (PCMs). For the sake of clarity, each technology and its functioning are described in a dedicated subsection where further information regarding properties, activation criteria, and results is provided. Moreover, all the results obtained simulating responsive technologies are compared with a law-compliant unresponsive reference model to understand potential benefits related to these systems.

For a better understanding of the phenomena under investigation, all the results obtained are shown, in the following sections, as percentage differences between the responsive models and a static reference model. Moreover, to consider the

Table 5.1 Simulation framework investigated

Application field	Technology	Climate scenarios	Locations	Exposures
Improvement of the thermal inertia	PCM embedded panel	TMY 2080	Larnaca Rome Helsinki	North East South West
Solar radiation control	Electrochromic	TMY 2080	Larnaca Rome Helsinki	East South West
Control of the heat flux	Variable U windows	TMY 2080	Larnaca Rome Helsinki	North East South West

different incidence of heating, cooling, and lighting in each climate, all the variations are reported as percentage variation of each energy source evaluated on the total consumptions as better described in Eqs. 5.1, 5.2, 5.3.

$$Heating\ variation = \frac{Heating_{Responsive} - Heating_{Reference}}{(Total\ consumptions)_{Reference}}[\%] \tag{5.1}$$

$$Cooling\ variation = \frac{Cooling_{Responsive} - Cooling_{Reference}}{(Total\ consumptions)_{Reference}}[\%] \tag{5.2}$$

$$Lighting\ variation = \frac{Lighting_{Responsive} - Lighting_{Reference}}{(Total\ consumptions)_{Reference}}[\%] \tag{5.3}$$

This approach allows to sum the percentage variations as they are all referred to the same common denominator (the total consumptions) simplifying the data interpretation. This common denominator depends on the selected technology to focus on the energy source actually affected by this technology. For example, PCMs and variable-U windows do not affect the lighting consumption, therefore the total consumptions are the sum of heating and cooling consumptions; on the contrary, the total consumptions for the EC windows also include the lighting consumption as it is affected by EC modulation. Hence, all the results provided in the next subsections are shown as stacked bars where each bar represents the percentage deviation of the total consumptions due to heating, cooling, and—if involved—lighting variations while the sum of all the bars depicts the overall percentage variation of the total consumptions compared with the reference model. Only in the last subsection, "Technologies comparison", the percentage variations are always referred to the [heating + cooling + lighting] total consumption—even though the technology does not affect the lighting—to allow clear comparisons between different technologies. Considering the structure of the equations, positive variations represent an increase of the total consumption in the responsive model and, vice versa, negative variations are overall reduction of the total consumption.

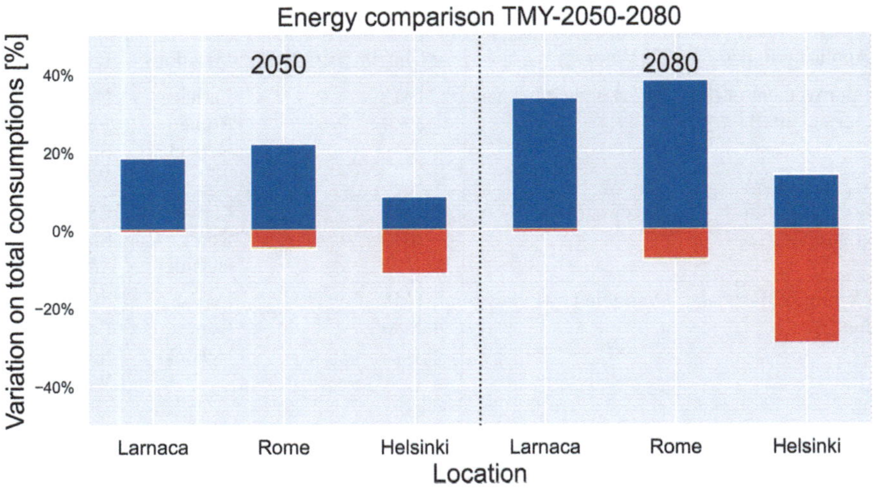

Fig. 5.4 Variations of the total consumptions [Heating (red) + Cooling (blue) + Lighting (yellow)] for a static model comparing TMY weather file with 2050 (left) and 2080 (right) weather file

Before moving to the description of the responsive technologies, a preliminary analysis regarding the incidence of future scenarios on consumptions was conducted to frame the topic. To that end, a static reference model was simulated in the two selected scenarios (TMY and 2080) and in an intermediate future projection (2050). Hence, the two future projections were compared with the typical weather file and their relative percentage variations were estimated according to Eqs. 5.4 and 5.5 and are summarised in Fig. 5.4.

$$Heating\ variation_{2050/2080} = \frac{Heating_{2050/2080} - Heating_{TMY}}{(Total\ consumptions)_{TMY}}[\%] \tag{5.4}$$

$$Cooling\ variation_{2050/2080} = \frac{Cooling_{2050/2080} - Cooling_{TMY}}{(Total\ consumptions)_{TMY}}[\%] \tag{5.5}$$

The results clearly show that the lighting variation is negligible while heating and cooling are strongly affected by these changes. In particular, the cooling consumption will likely rise in the next years increasing the total consumption up to 20% (in 2050) and 38% (2080) more than the typical weather scenario. On the other hand, the lower heating demand will decrease the total consumption in the 2050 and 2080 scenarios up to -11% and -30% respectively. Clearly, the location plays a pivotal role in these variations. Indeed, the heating incidence on total consumptions gradually increases moving towards colder climates. A similar trend could be expected also in the cooling consumptions but, in this case, the intermediate climate (Rome) registers the highest variations in both 2050 and 2080 scenarios. This trend can be easily explained looking at the air temperatures reported in Fig. 5.3a which shows a higher variation of the environmental parameters in Rome rather than in Larnaca.

Clearly, these results are all referred to static traditional technologies, therefore, they are dependent only on the climate change. It follows that they are fundamental starting point to understand the variations described in the next subsections which include both the effects of responsive systems and climate change.

5.1 Improvement of the Thermal Inertia: PCM Design

Considering the intrinsic control of PCMs, in this section the comparison is focused on two commercial PCM-enhanced plasterboard panels with same properties (thickness, density, and specific heat) but with transition temperatures slightly different (23 and 26 °C) and different melting latent heat (300 and 330 kJ/m^2). On the one hand, this choice leads to results pretty similar but, on the other hand, adding other commercial products with different properties (density, conductivity, specific heat) would have introduced too many variables that would have confused the results. The main properties of the commercial products selected and their relative ID codes—adopted through this chapter—are reported in Table 5.2. The use of passive strategies based on PCM in plasterboards allows to store a certain amount of latent heat in lightweight constructions that helps to reduce the surface temperature and to shift the peak load. When the PCM layer is adopted as the inner layer of the external walls it allows to stabilize the radiant temperature [1] while using it as external layer could mitigate the external temperature fluctuations. Therefore, in the considered examples, the two selected PCM plasterboards are located on both the inner (PCM_1 and PCM_2) and outer (PCM_3 and PCM_4) layers of the multilayered wall. Finally, all the PCM layers considered were implemented on all the façades of the building and simulated in all the selected locations for each climate scenario considered.

Regarding the comparison between responsive and reference models, they are characterised by the same multi-layered walls; nevertheless, the reference model is deprived of the temperature-enthalpy function representative of the change of phase. Consequently, the PCM layer is considered as a simple plasterboard in the reference model; this approach allows to consider only the responsiveness of the system as variable of the investigated problem.

Table 5.2 Input data adopted for the PCM embedded plasterboard model

ID code	Layer	Thickness [m]	Density [kg/m^3]	Specific heat [kJ/m^{2}K]	Melting latent heat [kJ/m^2]	Transition temperature [°C]
PCM_1	Inner	0.025	1000	28.3	300	23
PCM_2					330	26
PCM_3	Outer				300	23
PCM_4					330	26

The outcomes of the simulations conducted under the TMY weather file are summarised and reported in Fig. 5.5. As already mentioned, the plots represent the percentage variations of the total consumption—intended as heating + cooling in this case—caused by the variation of each energy demand source involved (heating and cooling for this technology). Before moving to the numerical results, it is worth reminding that all the results plotted in these sections are referred to the sum of all the thermal zones. Hence, null variations could be caused by conflicting results on different thermal zones characterised by different exposures rather than by inoperative technologies.

At first glance, it is clear that the effects of this technology on the total consumptions are limited, and this is mainly related to three different reasons. Firstly, the PCMs are adopted as fully passive technologies, in these examples, reducing the possible benefits connected to the forced convection on the panel. Secondly, PCMs are applied considering typical thicknesses for double panel plasterboards (2.5 cm) limiting, accordingly, the actual amount of PCM embedded in the wall. Thirdly, the same panels are applied, for sake of simplicity, on all the exposures without optimizing their position, melting temperature, and thicknesses. This last point is fundamental, in general, for responsive technologies which should be always optimized depending on exposures, location, and climate scenario to maximise the benefits and avoid counterproductive effects. For example, to maximise the benefits during both summer and winter periods, the multi-layered wall should be designed—for each location, climate, and exposures—considering different panels with specific targeted melting temperatures in different positions.

Going into more detail, the graph mainly shows uneven trends where:

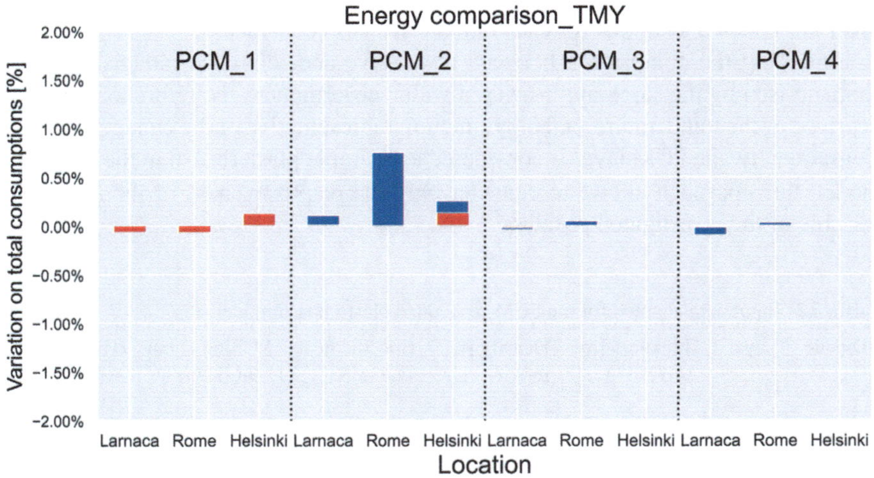

Fig. 5.5 Variations of the total consumptions [Heating (red) + Cooling (blue)] for the selected PCM models under TMY climate scenario compared to a static reference model

- the same material (PCM_1/PCM_3 and PCM_2/PCM_4) has opposite effects when applied on the inner or outer layer;
- the placement of a panel as inner layer (PCM_1/PCM_2) can affect mostly heating or cooling depending on the transition temperature and on the melting latent heat;
- the placement of a panel as outer layer (PCM_3/PCM_4) can affect mostly heating or cooling depending on the transition temperature and the melting latent heat;
- different climates do not affect unambiguously the consumptions which fluctuates between positive and negative variations depending on other design variables (material and position).

It follows that it is difficult to outline general trends for this technology due to the large number of variables (materials, position, climate, location, exposures) that can affect their behaviour. Furthermore, previous studies highlighted that the optimized placement of PCMs panels within the envelope depends on the thermal properties of the wall increasing the number of specific variables to account for a proper design of PCMs wall.

Figure 5.6 describes the outcomes of the same simulations run under the 2080 climate scenario to understand possible differences during the whole building life span. Firstly, when the PCM is implemented as inner layer of the wall (PCM_1/PCM_2) the percentage variations show trends similar to what already described for the current climate scenario but reduced in magnitude. Moreover, when the PCM is implemented as outer layer (PCM_3/PCM_4), the values are still lower than the TMY scenario, but, in this case, the variations fluctuate around zero and can be considered negligible.

Hence, in the light of the simulations conducted in both the scenarios, few main results can be outlined. Firstly, the effects of the climate change on the behaviour

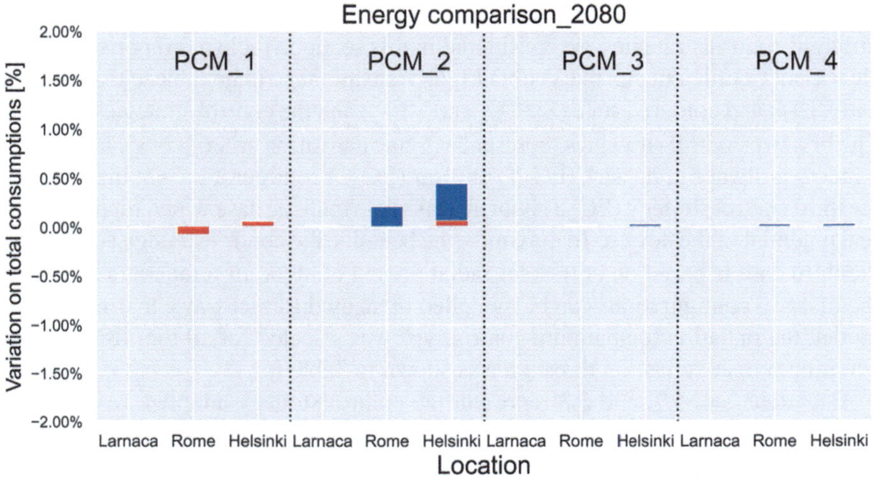

Fig. 5.6 Variations of the total consumptions [Heating (red) + Cooling (blue)] for the selected PCM models under 2080 climate scenario compared to a static reference model

of these systems are clear; for example, the benefits of adopting this material are zeroed in the future scenario when it is applied as outer layer, regardless of the transition temperature and melting latent heat. Instead, when it is adopted as inner layer, the effects are damped by the higher temperatures that could reduce the number of transitions of the PCMs. It follows that one of the limits of this technology is its intrinsic control that does not allow to adjust and tune the behaviour of the material in future scenarios according to possible variations of the boundary conditions (climate, building use, etc.). Therefore, it is fundamental to properly design these technologies considering the effects of possible future changes of the boundary conditions to avoid counterproductive effects during the building life span. To that end, a key role can be played by coupling different materials—with different transition temperatures—in different positions of the multi-layered wall depending on the exposure of each thermal zone.

Overall, despite its limited effects, only the first material applied as inner layer (PCM_1) can be considered as viable solution to reduce the energy consumption in hot and intermediate climates in both current and future climate scenarios. For this reason, only this solution will be considered in the comparison section (Sect. 5.4) as representative of this class.

5.2 Solar Radiation Control: Electrochromic Design

The analyses regarding the EC windows are based on an existing commercial product whose properties are reported in Table 5.3. A fundamental step for the modelling of an extrinsically controlled technology is the choice of the control strategy and its activation thresholds. To understand the effects of the modulation of the envelope properties—and the subsequent variation of the energy consumptions—three different control strategies are considered in this section. Table 5.3 reports the Solar Heat Gain Coefficient and the Visible Light Transmittance in both bleached ($SHGC_b$ and VLT_b) and coloured state ($SHGC_c$ and VLT_c) and the control strategies adopted. The first two control strategies represent extreme activation criteria; basically, the EC window is always in its dark (EC_1) or clear (EC_2) configuration. On the contrary, the third control strategy (EC_3) is an intermediate realistic case which represents the real potential of the device. To maximise the benefits, the control strategy adopted for the third case is based on both solar radiation and outdoor air temperature. Despite the first two configurations can be modelled in many different ways, to use the same model, the radiation/temperature control type was selected for all the cases, simply adopting proper activation thresholds as shown in Table 5.3.

Equations 5.6, 5.7, and 5.8 represent the control strategy adopted, respectively, for the EC_1, EC_2, and EC_3 cases: when the shading state is 0, the windows remain unchanged in their clear state, while, when the shading state is equal to 1, they automatically switch to their dark state.

Table 5.3 Input data adopted for the EC window model

ID code	$SHGC_b$ [−]	$SHGC_c$ [−]	VLT_b [−]	VLT_c [−]	Thresholds [°C] and [W/m²]
EC_1	0.47	0.06	0.63	0.01	−30 °C and 0 W/m²
EC_2					50 °C and 2000 W/m²
EC_3					15 °C and 120 W/m² and summer period

$$State_{EC\_1}(t) = \begin{cases} 0 \rightarrow G_f(t) < 0\frac{W}{m^2} \bigwedge T_o(t) < -30\,°C \\ 1 \rightarrow G_f(t) \geq 0\frac{W}{m^2} \bigwedge T_o(t) \geq -30\,°C \end{cases} \tag{5.6}$$

$$State_{EC\_2}(t) = \begin{cases} 0 \rightarrow G_f(t) < 2000\frac{W}{m^2} \bigwedge T_o(t) < 50\,°C \\ 1 \rightarrow G_f(t) \geq 2000\frac{W}{m^2} \bigwedge T_o(t) \geq 50\,°C \end{cases} \tag{5.7}$$

$$State_{EC\_3}(t) = \begin{cases} 0 \rightarrow G_f(t) < 120\frac{W}{m^2} \bigwedge T_o(t) < 15\,°C \vee t \notin summer \\ 1 \rightarrow G_f(t) \geq 120\frac{W}{m^2} \bigwedge T_o(t) \geq 15\,°C \bigwedge t \in summer \end{cases} \tag{5.8}$$

where t is the simulation timestep, G_f is the incident irradiance on the façade, and T_o is the outdoor dry bulb temperature. Considering that the control of the solar radiation in this study is assessed only from an energy perspective, all the control strategies allow to switch from clear to dark state only during the cooling period to maximise the winter benefits (as the thresholds could be exceeded also during winter). Despite glare issues could require darkening the EC device during winter, this topic is out of the scope of this analysis as it would add a further variable to the problem.

It is worth noting that, while the proposed realistic control strategy (EC_3) is quite common and effective, the thresholds adopted should be considered simply non-optimized average reference values derived from previous studies. Indeed, considering the nature of the responsive technologies, the activation thresholds should be optimized considering each location, climate scenario, and actual exposure of each window. For example, windows with different exposures are characterised by different incidence angles which could highly affect the value of the solar radiation thresholds. Nevertheless, for the purposes of this section, using the same control strategies ease the understanding of the energy phenomena involved simplifying the number of variables and the comparisons between different models.

All the simulated models were compared with a reference building with the same properties but characterised by law compliant static window (SHGC = 0.35 and VLT = 0.60) to highlight possible benefits related to the responsiveness of the device. Figure 5.7 sums up the results of the conducted analyses comparing reference and responsive models under TMY weather conditions. Firstly, the incidence of this technology on the energy consumption is more effective than those reported in the PCM section. Adopting this technology, the total consumption can be reduced or increased up to 15% simply because of the cooling or lighting variations. The graph clearly shows the greatest issue of this technology: the balance between cooling benefits and lighting drawbacks. Indeed, darkening the EC devices reduces the solar loads—and consequently the cooling demand—but also limits the incoming visible light, increasing the lighting consumption. This issue is mainly limited to the summer

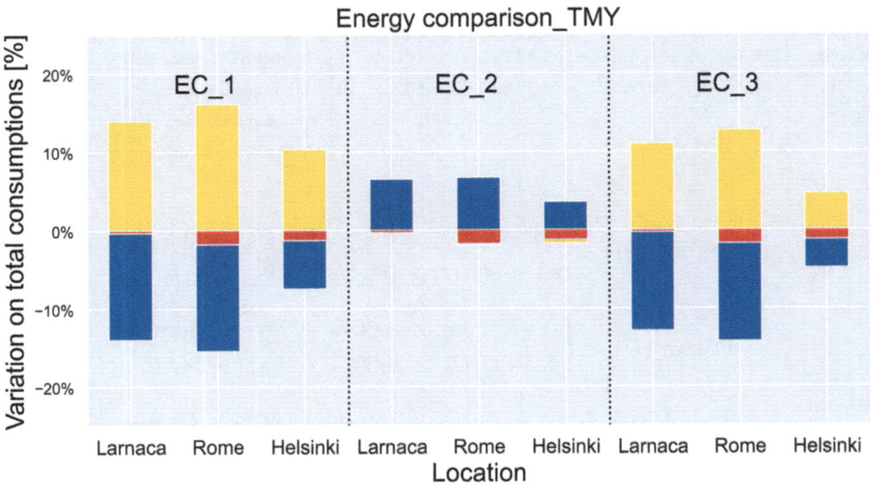

Fig. 5.7 Variations of the total consumptions [Heating (red)+ Cooling (blue)+ Lighting (yellow)] for the selected EC models under TMY climate scenario compared to a static reference model

period as, during winter, the main goal is the maximisation of the incoming solar radiation which can be easily pursued adopting proper seasonal activation schedule.

Looking more in detail the different activation criteria, all of them reduce the heating consumption thanks to the SHGC of the bleached state which is higher (0.47) compared to the law compliant reference value (0.35), increasing the solar gains during winter. Instead, considering lighting and cooling, EC_1 and EC_3 show similar trends while EC_2 consumptions are reversed. Considering the activation schedule that limits the glass darkening to the summer period, the differences between EC_1 and EC_2 regard the number of timesteps during summer when the device is on. Basically, in the EC_1 model the windows are always in their coloured configuration maximising both cooling benefits and lighting downsides. Nevertheless, if we limit the coloured state only when high solar radiation and outdoor temperatures occur (EC_3), the total consumptions increase of nearly $+1.5\%$ due to the cooling demand while the lower lighting demand drop by -4% the total consumptions (average value for different locations) leading to an overall energy payoff. These differences are larger for higher latitudes due to the different incidence angle of the solar beams. While, taking into account the always off activation criterion (EC_2), the windows are always in their clear state and, consequently, the cooling consumption increases in all the locations considered due to the higher SHGC. In the EC_2 case the larger cooling demand increases the total consumption of nearly $+ 5.6\%$ against a reduction related to lower lighting demand of only -0.2% on average.

Considering the total consumptions in the TMY climate scenario, adopting realistic activation thresholds (EC_3) leads to a reduction of the total energy consumption of -1.5% on average which is the best results obtained compared to the $+1.2\%$ and $+4\%$ registered for the EC_1 and EC_2 respectively. As already mentioned, the thresholds are not optimized for each location and exposure, thus, the potential benefits are higher than those reported in the graph which is mainly used to discuss the trends rather than the magnitude as actual potential variation. Moreover, the incidence of lighting in this specific model is particularly high due to its high illuminance setpoint (500 lx) while the incidence of cooling and heating benefits are limited thanks to the good EER and COP adopted. For example, simply reducing the setpoint to 300 lx would increase the total energy saving to −4% on average. Despite many studies consider only thermal consequences of implementing EC devices, these results highlight the key role of the lighting consumption to properly assess the real benefits of this technology. Furthermore, it should be noted that the EC windows in this case study are applied only to the 33% of total the wall surface; therefore, the potential of these devices on fully glazed buildings are extremely higher. Finally, the selected EC device is an already marketed product with very low VLT in its dark state; nevertheless, many products under development are extremely promising and could further improve the energy savings simply increasing the VLT of the dark state.

Moving to the future projections (Fig. 5.8), the EC devices show the same trend registered for the TMY weather scenario. Also for this technology, the variations are damped when compared with the TMY graph except for the EC_2 model which is worsen by the higher incidence of cooling demand. Specifically, considering the total consumption variations, the overall benefits in the EC_3 model are limited to −1.3% while the variations for the EC_1 and EC_2 models are +0.4% and +4.5% respectively.

Hence, these results confirm that EC technology can improve the overall energy behaviour of buildings during their whole life span if effectively managed. However, an additional advantage of these devices is that the reduction of benefits related

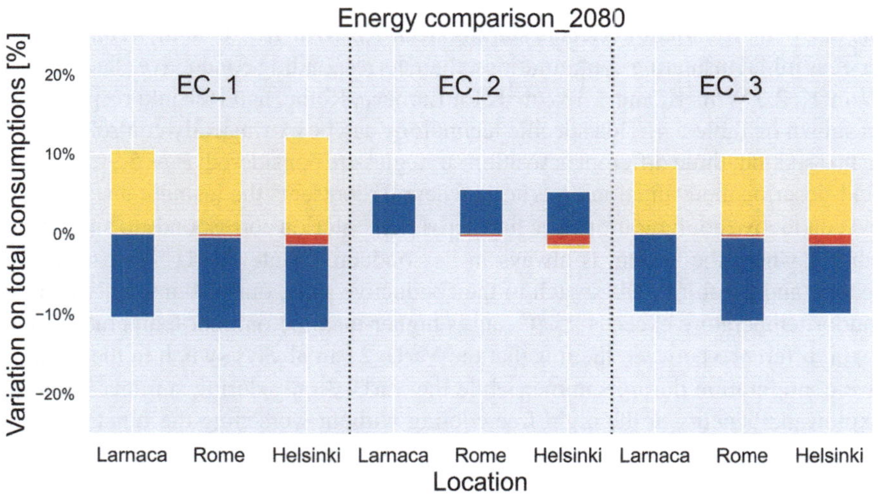

Fig. 5.8 Variations of the total consumptions [Heating (red) + Cooling (blue)+ Lighting (yellow)] for the selected EC models under 2080 climate scenario compared to a static reference model

to the changing climate scenarios can be overcome simply changing the activation thresholds and schedules during their life span according to the monitoring of the consumptions. Moreover, as better described in the last chapter, the flexibility of the extrinsic control allows to accommodate further possible changes related to urban context, building use, technology advancements etc.

Considering the outcomes obtained in both the climate scenarios, the EC_3 is the most performing control strategy; thus, it is selected as representative for this technology in the comparison subsection.

5.3 Control of the Heat Flux: Variable-U Windows Design

The last functioning criterion considered refers to the modulation of the heat fluxes through the building envelope. In particular, a variable-U window—based on the movable layer prototype proposed by Pflug [2]—was investigated in the same simulation framework described in the previous sections. This reference study [2] analyses the thermal transmittance of a double-glazed window with a low-density open pore melamine foam interposed in the window cavity. This translucent material can be moved in a middle or top configuration to, respectively, activate or prevent the convection around the panel as already explained in Chap. 3. Many configurations are assessed in the original study; among these, this book considers a double-glazed unit with a 10 mm top cavity and 17.5 mm lateral cavities. This specific model offers a modulation of the thermal transmittance of 0.53 W/m^2K (between 1.33 and 1.86 W/m^2K); considering the different locations selected, this range is applied to the law compliant U value for the following analyses. In particular, starting from the law reference value, adopted as insulating state, the conductive state is obtained subtracting the 0.5 W/m^2K variation. Hence, starting from 2.2 W/m^2K, 1.8 W/m^2K, and 0.8 W/m^2K as initial insulating configurations, their corresponding conductive states are 2.7 W/m^2K, 2.3 W/m^2K, and 1.3 W/m^2K for Larnaca, Rome, and Helsinki respectively, as shown in Table 5.4. This specific technology can be extrinsically controlled, and, in this section, three different activation strategies are considered; Eqs. 5.9, 5.10, and 5.11 describe their functioning criteria where 0 represents the insulating state and 1 the conductive configuration. The first activation criterion corresponds to an extreme control where the system is always in its conductive state (VarU_1). Instead, the second and the third ones switch to the conductive state, during summer, when the indoor temperature exceeds 25 °C and is higher than the outdoor temperature. The main difference between them is that the VarU_2 can always switch to the conductive configuration during summer while the VarU_3 only during summer nights to exploit the benefits of the night free cooling without worsening the functioning of the building systems—considering the envelope thermal inertia—when the cooling is on.

$$State_{VarU\_1}(t) = \begin{cases} 0 \rightarrow & \mathrm{T}_{in}(t) < 0\,^{\circ}\mathrm{C} \\ 1 \rightarrow & \mathrm{T}_{in}(t) \geq 0\,^{\circ}\mathrm{C} \end{cases} \tag{5.9}$$

$$State_{VarU\_2}(t) = \begin{cases} 0 \rightarrow T_{in}(t) > T_o(t) \vee T_{in}(t) > 25\,°\mathrm{C} \vee t \notin summer \\ 1 \rightarrow T_{in}(t) \leq T_o(t) \bigwedge T_{in}(t) \leq 25\,°\mathrm{C} \bigwedge t \in summer \end{cases} \tag{5.10}$$

$$State_{VarU\_3}(t) = \begin{cases} 0 \rightarrow T_{in}(t) > T_o(t) \vee T_{in}(t) > 25\,°\mathrm{C} \vee t \notin summer\ night \\ 1 \rightarrow T_{in}(t) \leq T_o(t) \bigwedge T_{in}(t) \leq 25\,°\mathrm{C} \bigwedge t \in summer\ night \end{cases} \tag{5.11}$$

Using 25 °C as temperature threshold allows to avoid heat losses while the cooling system is on as the cooling setpoint is 24 °C; instead, adding the nighttime control in the third strategy aims to reduce the possible heat losses during the day between the switch on and switch off of the cooling system due to the transient nature of thermal phenomena.

Starting from these inputs, the models were simulated under TMY and 2080 climate scenarios and compared with a reference static model characterised by the same windows fixed in their insulating state. Similarly to what already described for the PCMs, in this case the analyses are focused only on the heating and cooling consumptions which are involved in this case. Figure 5.9 shows the results for the typical climate scenario where the difference between the first and the last two control strategies is clearly reflected in terms of energy consumption. When the window is always in its conductive state (VarU_1), the higher heating demand increases the total consumptions—proportionally with the HDD—ranging from +0.7% up to +5%.

Table 5.4 Input data adopted for the variable-U walls model

ID code	Conductive state thermal transmittance [W/m^{2}K]	Insulating state thermal transmittance [W/m^{2}K]	Thresholds
Larnaca			
VarU_1	2.7	2.2	(T_{in} > 0 °C)
VarU_2			(T_{in} > T_o) and (T_{in} > 25 °C) and (Summer period)
VarU_3			(T_{in} > T_o) and (T_{in} > 25 °C) and (Summer nights)
Rome			
VarU_1	2.3	1.8	(T_{in} > 0 °C)
VarU_2			(T_{in} > T_o) and (T_{in} > 25 °C) and (Summer period)
VarU_3			(T_{in} > T_o) and (T_{in} > 25 °C) and (Summer nights)
Helsinki			
VarU_1	1.3	0.8	(T_{in} > 0 °C)
VarU_2			(T_{in} > T_o) and (T_{in} > 25 °C) and (Summer period)
VarU_3			(T_{in} > T_o) and (T_{in} > 25 °C) and (Summer nights)

Furthermore, the reduction of the cooling consumption leads to a slight increase of the total consumptions in the hottest climate (Larnaca + 0.1%), and a limited reduction in the other locations (−1% for Rome and −1.3% for Helsinki). Clearly, overall, this control strategy cannot be considered effective due to the winter behaviour and to the higher heat exchange during summer when the cooling system is on. For the hottest climate, the energy loss when the cooling is on actually overcome the benefits of the free cooling related to the higher thermal transmittance of the windows. On the contrary, the more responsive control strategies VarU_2 and VarU_3 lead to lower cooling demands that reduce the total consumptions of −0.35% and −0.31% on average. Despite the values are quite similar, limiting the conductive state only to summer nighttime slightly reduces the overall benefits; comparing the cooling variations of the three models it is clear that the control strategies can be further improved. Indeed, in the VarU_1 model, except for Larnaca, the cooling benefits are higher notwithstanding the energy loss during the functioning of the cooling system. It follows that the optimal activation threshold is—from a cooling point of view—something in between the first and second model; hence, finding this optimal control strategy could further improve the maximum cooling savings registered in these simulations. Moreover, as already stated for other technologies, the control strategy in these simulations is the same for the whole building; therefore, further improvements can be achieved simply reasoning on each zone separately. Another important variable to be considered to assess the overall benefits of these devices is the Windows to Wall Ratio (WWR); as already stated for the EC windows, the energy savings could be maximised applying these systems on highly glazed buildings.

The importance of finding an optimized control for each considered case is highlighted in Fig. 5.10 where the effects of the climate change on the previous control

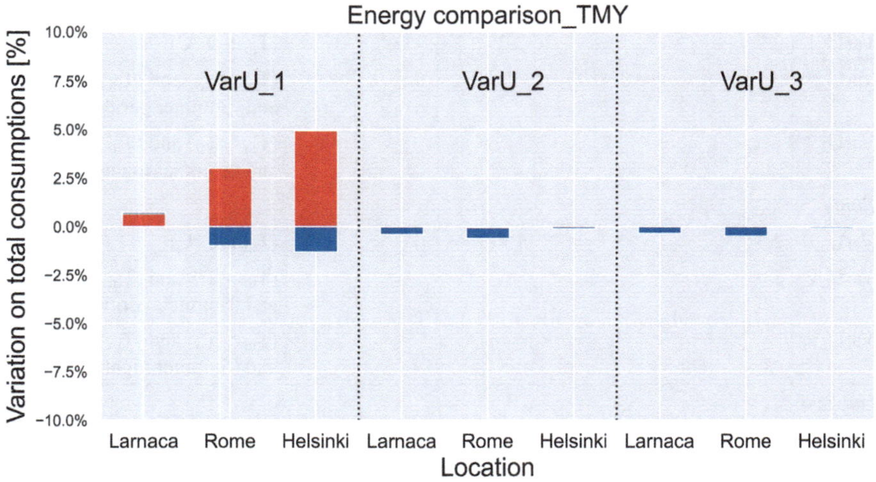

Fig. 5.9 Variations of the total consumptions [Heating (red) + Cooling (blue)] for the selected variable-U models under TMY climate scenario compared to a static reference model

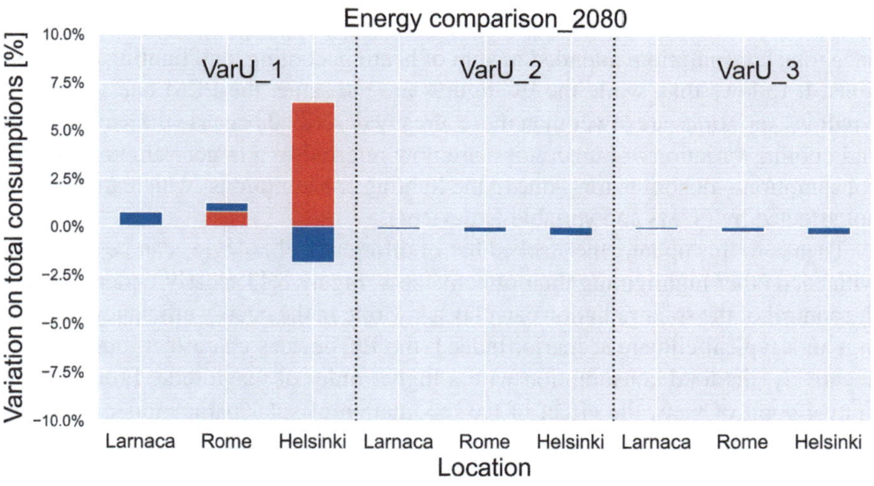

Fig. 5.10 Variations of the total consumptions [Heating (red) + Cooling (blue)] for the selected variable-U models under 2080 climate scenario compared to a static reference model

strategy are depicted. With reference to the VarU_1 model, leaving aside the heating consumptions, the increase of the outdoor temperature increases the number of timesteps when the cooling is on, maximising the drawbacks of this control strategy. Indeed, in the 2080 scenario, both Larnaca and Rome increase their total consumptions due to the higher cooling demand while only Helsinki improve the summer benefits.

Moving to the VarU_2 and VarU_3 models, the trends are similar to what already discussed for the TMY case; nevertheless, the benefits are lower for Larnaca and Rome while Helsinki is the only location where higher savings are registered thanks to an increase of the cooling incidence caused by higher outdoor temperatures. Unlike EC devices, the variation of the boundary conditions related to the climate change can only be partly solved updating the control strategy as the activation thresholds adopted in this section are already set by the cooling setpoint and the indoor/outdoor air temperatures limiting the possibility of variations.

According to the results discussed, the overall best configuration is the VarU_2 which will be considered in the next section for the technology comparison.

5.4 Technologies Comparison

Finally, a comparison between the different technologies considered in the previous subsections is conducted to identify possible hints for the design of these devices. This assessment is conducted considering the best configurations emerged during the first exploratory simulations. Nevertheless, in order to make the comparison reliable, the results obtained from the simulations are reported referred to the same reference total consumption, intended as sum of heating, cooling, and lighting consumptions. It follows that, while the EC figures are the same, the PCM and variable-U windows variations are lower than those already described because the same heating and cooling variations—numerator—are now referred to a larger value of the total consumption—denominator—due to the lighting consumptions (which are however not affected by PCMs and variable-U devices).

Thanks to this update, the stacked bar of different technologies can be compared with each other highlighting their effectiveness. Figure 5.11 clearly underlines how the control of the solar radiation can play a key role in the energy efficiency of buildings in a typical climate scenario. Indeed, the EC devices can affect, positively or negatively, the total consumption with a higher order of magnitude. From a quantitative point of view, the effect of the modulation of solar parameters can change the lighting and thermal consumptions varying the total consumptions up to nearly ±15%. On the contrary, the control of the heat flux and the improvement of the thermal mass can affect the total consumption (including the lighting) a maximum of 0.3% and 0.1% respectively. On the one hand, the higher incidence of EC on the total consumption recognizes the potential of these devices; nonetheless, on the other hand, their wrong design can significantly increase the energy consumption and the users' discomfort.

The effects of the climate change on all the technologies analysed are reported in Fig. 5.12 where the same trend described for the TMY climate scenario can be identified. While the EC devices are still the most effective, their incidence is reduced (up to nearly ±10%) in future climates and the differences between warm

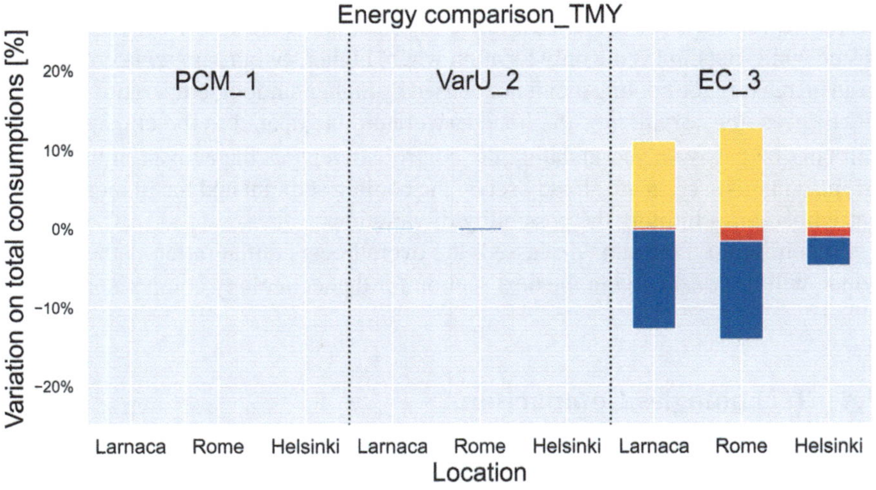

Fig. 5.11 Variations of the total consumptions [Heating (red) + Cooling (blue) + Lighting (yellow)] for the PCM_1, VarU_2, and EC_3 models under TMY climate scenario compared to a static reference model

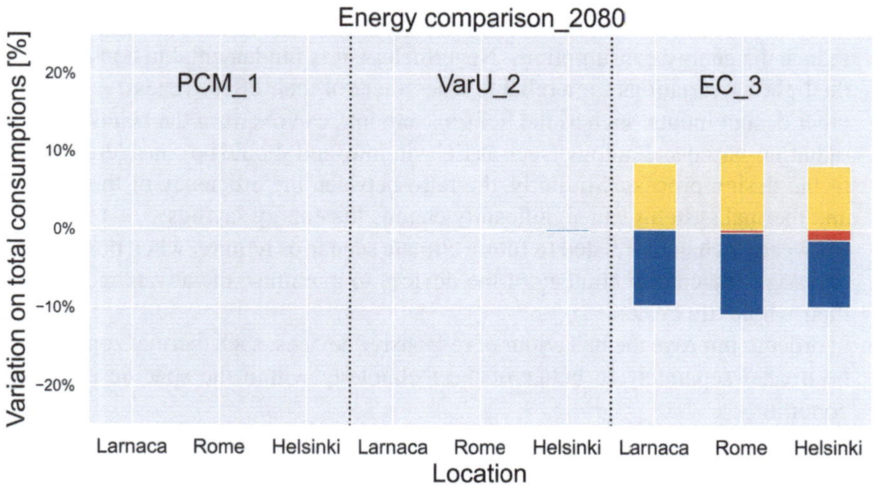

Fig. 5.12 Variations of the total consumptions [Heating (red) + Cooling (blue)+ Lighting (yellow)] for the PCM_1, VarU_2, and EC_3 models under 2080 climate scenario compared to a static reference model

and cold locations are minimized. As already mentioned, the control strategy for extrinsically controlled devices can and should be updated in future climate scenarios; in this hypothesis, the EC potentials would be higher in the 2080 scenario. This point emphasizes the importance and flexibility of the extrinsic control that allows to tune and refine the responsive behaviour of the device in progress, where possible, in accordance with monitored data.

Another important outcome that can be provided by the conducted analyses is the strong dependency on the boundary conditions. Climate, locations, exposure, setpoints, building type, WWR, systems efficiency, etc. are—along with the activation criterion—the core of the functioning of responsive devices. For clarity and explanatory reasons, the applications conducted in this chapter were simplified treating all the thermal zones and climate scenarios analogously. Nevertheless, to exploit the actual potential of these technologies, the control strategy or the design of the system itself should be optimized for each application. Hence, for example, EC windows applied on different exposures or in different locations could be characterised by different irradiance thresholds; PCMs could be characterised by different transition temperatures depending on the specific thermal zones.

To sum up, it is extremely hard to define stiff cases and designing rules for responsive technologies due to their specific nature. The main goal of this section was to bring out all the possible considerations regarding the choice and the control of a responsive device—in current and future scenarios—to approach its design mindfully. The following list aims to summarize these aspects accordingly with the results obtained by the conducted simulations:

- the control of the solar radiation seems to be the most effective approach to reduce the energy consumptions. Nevertheless, it is fundamental to consider also the lighting variations for a reliable assessment of their effectiveness;
- other design inputs, such as the lighting setpoint, can overturn the behaviour of a radiation-based technology (such as EC window) and should be considered as part of the design process. Similarly, the ratio between the efficiency of the lighting and thermal systems can significantly change the energy savings;
- the weather changes related to future climate scenarios require, when possible, an update of the control strategy of the devices to maximise the advantages during their whole life cycle;
- in order to improve the behaviour of responsive devices, each thermal zone should be treated separately to better fit the technology within the specific boundary conditions;
- extrinsically controlled systems offer higher flexibility and allow to adjust the behaviour of the device adapting it to changing environments (climate change, building type, etc.);
- intrinsically controlled systems should be properly designed to avoid counterproductive effects. In particular, PCMs should be studied in different positions of the multi-layered wall and with different transition temperatures to find the right balance between winter and summer behaviour.

References

1. F. Fiorito, Phase-change materials for indoor comfort improvement in lightweight buildings. A parametric analysis for Australian climates. Energy Proc. **57**, 2014–2022 (2014). https://doi.org/10.1016/j.egypro.2014.10.066
2. T. Pflug, N. Nestle, T.E. Kuhn, M. Siroux, C. Maurer, Modeling of facade elements with switchable U-value. Energy Build. **164**, 1–13 (2018). https://doi.org/10.1016/j.enbuild.2017.12.044

Chapter 6
Evaluating the Building Resilience: Energy Flexibility Index

Abstract The main aim of this chapter is to investigate the building energy simulations from a wider perspective to take into account several future uncertainty sources. For example, climate change, technological advancements, variations of the urban context and building use can play a pivotal role—for both responsive and static technologies—in the future energy consumption of our buildings, overturning their current energy behaviour. Starting from a validated study, this chapter reports an innovative metric that summarises—in a single index called Energy Flexibility Index—the behaviour of a specific technology in a series of possible future scenarios.

Keywords Responsive envelope · Resilience · Future scenarios · Climate change · Architectural engineering

Previous chapters describe the theoretical background and practical hints required to design responsive envelopes considering the environmental variations related to the climate change. Undoubtedly, considering their impact on our lives, the climate change and the climate resilience will be among the most discussed topics in the next years. Nevertheless, currently, few approaches are available to frame these themes in a shared and well-structured methodology that allows to investigate the problem in a comprehensive way. The results described in the fifth chapter highlighted how strong is the dependency of the energy behaviour of a responsive building on specific environmental, technological, and constructive factors. Simply changing exposures, climate scenarios, control strategies, building systems, etc. drastically changes the energy behaviour and the number of simulations and outcomes to be handled. This concept can be applied on both responsive and static technologies; nevertheless, the former—due to their "interactive" behaviour—are more unpredictable and could lead to particularly unexpected performances. On the one hand, considering different options is doubtless an advisable approach but, on the other hand, it increases the number of the results that would be hardly manageable and comparable. Moreover, during the design phase or in the scientific research, it could be difficult to clearly define the boundary conditions where the designed building will operate for the following fifty years. Indeed, when it comes to future assessments, many

F. Carlucci et al., *Responsive Envelopes and Climate Change*, Digital Innovations in Architecture, Engineering and Construction,
https://doi.org/10.1007/978-3-031-58101-4_6

boundary conditions can be affected by strong uncertainty—which can be only partly predicted—that can also influence the envelope design.

Currently, the future-projected analyses are beginning to be conducted mainly focusing the attention on a single future variable—mostly, the climate change—neglecting many other uncertainty sources that could have significant impacts on the envelope design. Looking at the trends registered in the past years, building use [1], internal loads [2], social and demographic fluctuations [3], and urban context [4, 5] are examples of boundary conditions that have significantly changed over the years. For example, the current lighting consumption is significantly lower than twenty years ago thanks to the spread of LED lamps. This sudden reduction of the lighting consumption has changed the balance between cooling and lighting; indeed, thanks to the lower energy incidence of the artificial lighting systems, reducing the amount of sunlight—for example, by using a summer shading device—could be more convenient than in the past as it strongly decreases the cooling demand to the detriment of a slight increase of lighting consumptions.

In general, studying several possible scenarios allows to design flexible and versatile buildings and in other fields this approach is already widespread. For example, in architectural design, the analysis of multiple scenarios is often considered in the design of the spatial layout of buildings to account for possible future variations of the boundary conditions (number of occupants, tenants, building use, etc.). Designing buildings that can adapt themselves in a versatile manner to unpredictable future events also from an energy point of view should be a goal to be pursued in the next years.

The responsive envelopes can play a pivotal role in achieving this goal, also in a long-term scenario, thanks to their capability to change their properties according to external stimuli. Though, due to their lower predictability, it is fundamental to proper design the possible operating scenarios to avoid counterproductive effects. Nevertheless, moving from a single variable projection to a wider future analysis means increasing the number of possible combinations and the number of the results to be analysed. Consequently, the intrinsic complexity of managing several scenarios and multiple results are the main reasons that slow down the spread of these analyses.

Starting from these premises, this section, aims to provide a first introduction to a new approach that considers the great influence of different uncertainty sources in the building energy assessment. According to the methodology proposed in a dedicated study [6] this chapter describes a new way to account for many different uncertainty sources using a single indicator: the Energy Flexibility Index. This approach is not univocally related to specific adaptive systems, but it could be also applied to compare different static technologies; however, in this book, the main focus regards responsive technologies. Following subsections firstly describe few of the alternative future scenarios considered by the authors in order to define a matrix that summarises the possible variations of the boundary conditions in the future. Hence, a detailed description of the proposed index and the interpretation of the results is provided as reference for future studies.

6.1 Possible Future Scenarios

Every building energy analysis is strictly dependant on a series of specific assumptions that define the simulation framework of a specific model. Usually, these assumptions are based on fixed typical or standardised input data (typical weather files, standardized occupancy, load profiles, etc.) which may become inconsistent when we project into the future our assessments, considering the whole building life. To switch from a typical to a future-oriented approach, it is fundamental to be aware of the main possible changes that could occur during the building service life.

According to these premises, the future of the designed building might include one or more alternative possible scenarios; potentially, every combination of changes of the boundary conditions could take place during the building service life cycle. Nevertheless, from a realistic perspective, not every scenario has the same impact on the performance nor the same probability. Therefore, to avoid overestimating the number of possible scenarios, it is important to consider only those that are characterised by high impact and high probability. Figure 6.1 schematically represents the different impact of two possible scenarios (the climate change and the widespread of sustainable materials); in this case, it is clear that despite both the drivers are characterised by high probabilities, only the climate change has a direct higher impact on the building energy consumption. Similarly, also the spread of sustainable transportation can be neglected—in the building energy field—when compared to climate, architectural, or urban changes as it could provide slight microclimate improvements but the overall impact of sustainable transportation on buildings' consumptions will be probably limited. Next paragraphs sum up the most impactful and likely scenarios according to the authors' opinion which will be considered to define the new index in the next subchapter.

Accordingly to what already stated in the previous chapters, the widest shared future storyline regards the weather, which is considered in this book as main future scenario. Thanks to a new awareness of the risks connected to the climate change,

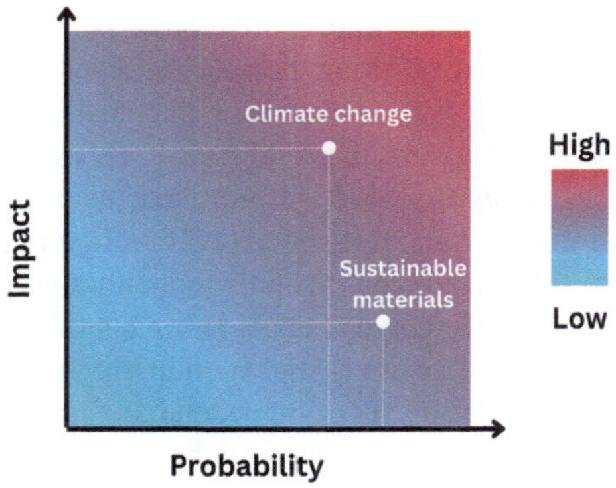

Fig. 6.1 Example of the incidence of a specific future scenario related to its impact-probability assessment

this scenario is getting more consensus and spread in the scientific field. Despite this, currently, the energy analyses are mainly conducted using the Typical Meteorological Year (TMY) weather files [7] or one of its following versions (TMY2 released in 1995 or TMY3 released in 2008 [8, 9]). These weather files are referred to data collected in at least 10 years, merged in a single file considering the most typical and representative trends. Nonetheless, many other future variations—regarding less common scenarios—can have significant impacts on the overall energy behaviour of the buildings. Following paragraphs provide a brief description of the main "secondary" alternative scenarios according to the authors' opinion.

The last decades have clearly demonstrated a reduction trend in the energy consumption of electrical equipment and lighting. This trend is translated into a reduction of the internal gains which are, as a result, usually overestimated in the typical energy simulations. Overestimating the internal gains means overrating the cooling demand and underestimating the heating consumption. An interesting study [10] analysed the variations of the internal gains in office buildings in the last years highlighting a downward trend of the power absorption—and of the internal gains—since 2000. As already mentioned, this reduction is driven by the spread of LED lamps that reduces of nearly 50% the lighting loads. Considering the rising interest in energy efficiency and the sustainable goals set, it is reliable considering this trend confirmed also in the next years.

Another minor future scenario, which is rarely considered in the energy field, regards the urban context. Realising or demolishing a building or a green area can have many effects on the surroundings which can be reflected on the energy behaviour of the designed building. The mutual shading, the urban canyon effect, and the urban heat island effect are examples of environmental variations—related to the local microclimate—triggered by changes in the urban topology. Moreover, the frequency of urban changes seems to be higher in the last years as underlined in a specific study [11] regarding the buildings' lifespan which could be shorter than expected, mainly considering office buildings. The authors investigate the average lifespan of recent buildings, and the obtained outcomes highlight shorter lifespans for office buildings (nearly 40 years) compared to residential buildings (nearly 170 years). It follows that, especially multifamily housings, could face several changes in the surrounding urban context during their service lives which could affect their energy behaviour.

Finally, the last important "secondary" future scenario considered in the reference study [6] regards the building use. This scenario is going to become more likely in the next years as proved by several recent examples where many building uses—ranging from office and residential [12] to religious [13] and commercial [14]—were adapted to fulfil new needs. This is probably the most effective and likely change that can overturn the energy behaviour of a building. Indeed, during the building service life, the socio-economic context often triggers changes in the building use driven by many factors such as, for example, market forces or new working or living approaches [15]. Another important driver for this secondary future scenario is the pursuit of circularity in the building sector; this approach leads to prefer the reuse of buildings to avoid the emissions related to transportation and reprocessing of the building material [16]. While this new trend has started a new way to design architecturally flexible layouts, the energy flexibility of this approach is often not considered. Indeed, despite the reduction of the embodied carbon footprint, changing the building use could reduce the energy efficiency of the building during its whole life span [17] if not properly designed.

This brief overview hints that the real future scenario could include one or more variations of the boundary conditions depending on several uncertainty sources. Hence, rather than considering a single scenario, a matrix of possible future scenarios can be defined to conduct future-oriented simulations. Considering the most recent IPCC climate change report [18], it is highly likely that global warming will exceed 1.5 °C during the twenty-first century and it is difficult to limit it below 2 °C. Therefore, the climate change is considered, in these future simulations, as "main" change; it follows that in every considered future scenario future climate projections will be considered instead of current typical weather files. Figure 6.2 summarises this concept in a sample matrix where the climate change (C) is accounted in every future scenario, considering two possible alternatives (C_1, C_2) that represent respectively the projection of the current weather file (C_0) in 2050 (C_1) and 2080 (C_2) respectively. On the contrary, additional secondary variables—urban context (U), building use (T), and internal loads (L)—could confirm their current figures (U_0, T_0, L_0) or could change in the future (U_1, U_2, T_1, L_1 in this example) according to specific authors' considerations. Considering this example, the baseline is the current scenario ($C_0.U_0.T_0.L_0$) where all the variables are considered with their current or standard values; instead, regarding the future scenarios, 9 different combinations ($U_x.T_y.L_z$) can be considered for each climate projection (C_1 and C_2) for a total of 19 simulations, including the baseline.

These theoretical premises outline a series of possible alternative scenarios, and their combinations, that can significantly affect the contexts considered during the design phase. These changes can improve or worsen the energy behaviour of buildings, and responsive envelopes could be the answer to these fluctuations of the boundary conditions, if properly designed. Nevertheless, managing simultaneously all these uncertainty sources and simulations can be a challenging goal. In the above example, simply considering two climate projections, two alternative urban contexts, one alternative building use, and one alternative internal loads scenario means managing 19 different scenarios (1 baseline and 18 future projections) and, consequently, results. The following section provides the methodology to summarise the results of the considered matrix in a single index to simplify the energy assessments and the comparison of different envelope technologies.

6.2 A Comprehensive Index: Energy Flexibility Index

Considering the number of possible future scenarios, the analysed component, control strategy, or building should be tested in all these contexts to assess its overall behaviour during the whole service life. To summarise the flexibility of a component or technology in a single index—called Energy Flexibility Index (EFI)—, the

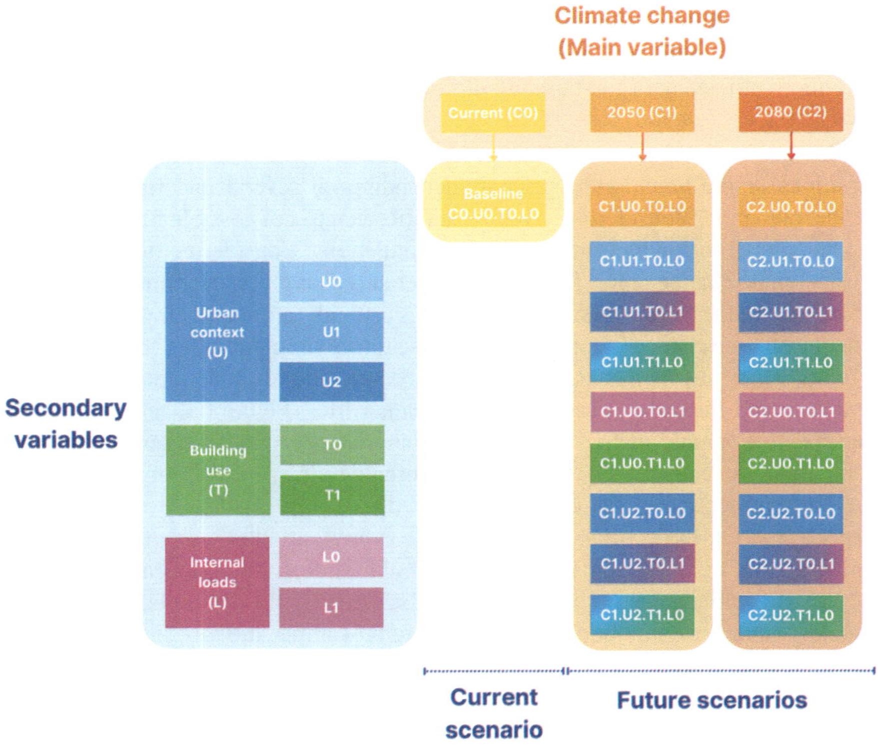

Fig. 6.2 Combination of the possible future scenarios considering two alternative climate projections (C1, C2), two additional possible urban contexts (U1, U2), one alternative building use (T1), and one alternative internal loads scenario (L1). The index 0 represents the current or typical scenarios

reference study [6] suggests, firstly, to simulate every scenario deducted from the future matrix. Despite this approach could be adopted and adapted to different applications (components, control strategies, technologies, building, etc.), for sake of coherence—according to the reference study—the flexibility of a specific control strategy is assessed in this chapter with reference to the future matrix reported in the previous section. In particular, in this example three different control strategies are considered to compare their different energy flexibility.

Once all the simulations are completed, a key stage of the methodology is the comparison between the different future elements of the matrix (18 in this example) and the baseline (1 for all the cases considered) for each control strategy considered (for example three for this case study). It follows that each control strategy is simulated in 19 (18 future and 1 baseline) different contexts, for a total number of 57 simulations. This comparison is conducted evaluating the Energy Consumption Variation (ECV), defined according to Eq. 6.1. For each kth control strategy and for each ith future scenario (S_i) the ECV_{ki} is defined as the difference between the energy consumption in the ith future scenario ($EC_{k\,Si}$) and the current scenario baseline ($EC_{k\,CS}$) divided by the current scenario energy consumption.

$$ECV_{ki} = \frac{EC_{kSi} - EC_{kCS}}{EC_{kCS}} \tag{6.1}$$

This ratio basically represents the relative variation of the energy consumption in each future scenario (S_i) with respect to current scenario (CS) baseline; hence, negative values of the ECV represent future reduction of the energy consumptions (improved performances) while positive figures represent an increase of the consumptions in the future scenario (reduced performances).

According to this definition, for each control strategy (3 in this example) and for each future scenario (i.e. 18) we can calculate the ECV_{ki} ($18 \times 3 = 54$ values in this example). Nevertheless, each ECV_{ki} is referred to a specific baseline ($Ec_{k\,CS}$) which depends on the kth control strategy considered; therefore, currently, comparing these values cannot be considered reliable. For example, a static low-insulated envelope simulated in a very cold climate (high $EC_{k\,CS}$) could improve its overall energy behaviour thanks to a strong reduction of the heating demand and a very slight increase of the cooling consumption in the future climate scenario due to the climate change effect (low $EC_{k\,Si}$). On the contrary, an optimised responsive device could already perform better in the current scenario (low $EC_{k\,CS}$) reducing, as consequence, the ECV value. It follows that comparing these two options, the ECV of the static component could be higher than the responsive device; nevertheless, this difference is simply related to a worst behaviour of the static technology in the baseline current scenario rather than on a real higher flexibility of the component. In order to account for this dependency, the reference study proposes the adoption of a weighting factor w to normalize the energy consumption variation with respect to a reference technology in the current scenario (ref_{CS}). The reference technology can be, for example, a traditional law-compliant static envelope and is required as reference to compare the behaviour in the baseline scenario of the different options considered in the flexibility assessment. Hence, the weighting factor (w) can be calculated—for each kth control strategy—starting from the ratio between the energy consumption of the reference technology in the current scenario ($EC_{ref\,CS}$) and the energy consumption of considered kth control strategy ($EC_{k\,CS}$) as described in Eq. 6.2. Considering that the ECVs can assume both negative (improved performance) and positive (reduced performance) values, the weighting factor can assume values between 0 and 1 or greater than 1 to amplify or reduce the benefits or the drawbacks of a specific scenario depending on the ECV sign. To that end, the exponent of the ratio assumes only $+1$ or -1 values, depending on the sign of the ECV. Indeed, when the $EC_{k\,Si}$ is higher than $EC_{k\,CS}$ the ECV is positive while the exponent is negative and vice versa. This formulation allows to:

- increase the weighting factor ($w > 1$) if the kth control strategy technology performs better (lower consumption) than the reference technology ($ratio > 1$) and the ECV is negative (hence the exponent is positive);

- increase the weighting factor ($w > 1$) if the kth control strategy technology performs worse (higher consumption) than the reference technology (ratio < 1) and the ECV is positive (hence the exponent is negative);
- decrease the weighting factor ($0 < w < 1$) if the kth control strategy technology performs better (lower consumption) than the reference technology (ratio > 1) and the ECV is positive (hence the exponent is negative);
- decrease the weighting factor ($0 < w < 1$) if the kth control strategy technology performs worse (higher consumption) than the reference technology (ratio < 1) and the ECV is negative (hence the exponent is positive).

$$w = \left(\frac{EC_{refCS}}{EC_{kCS}}\right)^{-\frac{EC_{kSi} - EC_{kCS}}{\left|EC_{kSi} - EC_{kCS}\right|}} \tag{6.2}$$

It follows that the new indicator adopted to compare the energy variation of the kth control strategy in the ith scenario is the Normalized Energy Consumption Variation ($NECV_{ki}$) which is simply the product between the ECV_{ki} and the weighting factor was described in Eq. 6.3.

$$NECV_{ki} = ECV_{ki} * w \tag{6.3}$$

In this way, when the ECV is negative (lower consumption in future scenarios) if the kth control strategy performs better than the reference technology the weighting factor amplifies ($w > 1$) the benefits while, if the reference technology performs better than the control strategy, the weighting factor reduces ($0 < w < 1$) the benefits. On the contrary, when the ECV is positive (higher consumption in future scenarios) if the kth control strategy performs better than the reference technology the weighting factor reduces ($0 < w < 1$) the disadvantages while, if the reference technology performs better than the control strategy the weighting factor amplifies ($w > 1$) the disadvantages. The most likely scenario is that the analysed technology performs better than the reference one; nevertheless, this metric allows to properly assign different weights to account for every possible case.

Therefore, considering the kth control strategy, the NECV allows to directly compare all the elements of the future matrix in a single index called EFI_k. As highlighted by the subscript k, the EFI is not referred in general to all the control strategies evaluated but it summarises the behaviour of each specific control strategy (or in general technology) in all the considered future scenarios. From a mathematical point of view, the EFI is simply the opposite of the mean referred to all the n future elements of the matrix ($n = 18$ in this example) of all the NECV for a specific control strategy as described in Eq. 6.4. In order to have direct proportionality between the EFI and the reduction of the energy consumption, the authors consider the opposite of the mean; it follows that higher values of EFI correspond to lower average NECVs and, hence, to better performances and flexibility.

Nevertheless, in all the applications, the mean is statistically significant only if it is representative of the data considered. Hence, before calculating the EFI, the reference study suggests adopting the Coefficient of Variation (CV) as indicator to verify the NECV dispersion. When the CV—which is the ratio between the standard deviation σ and the mean μ—is lower than 0.5, the EFI can be calculated and can be considered representative of the analysed matrix. On the contrary, if the CV exceeds this threshold, the flexibility is not symmetric, hence, there is one or more alternative scenarios that are strongly different from the others. In this case, the authors suggest splitting the analysis by isolating the outliers—usually the change of building use—in a separated EFI analysis in order to limit the CV value and obtain representative data.

$$EFI_k = \begin{cases} EFI_k = -\frac{1}{n} * \sum NECV_{ki} \rightarrow (CV = \frac{\sigma}{\mu}) < 0.5 \\ \\ Asymmetric\,flexibility \rightarrow (CV = \frac{\sigma}{\mu}) > 0.5 \end{cases} \tag{6.4}$$

Finally, the authors also propose a further indicator to better describe the behaviour of the technology in all the considered future contexts. In particular, despite a single index ease the management of the results, few details could be missed by averaging all the scenarios. To that end, a classification based on the statistical analysis of the NECV obtained is suggested to understand the distribution around the zero of the results as shown in Fig. 6.3. Depending on the distribution of the different quartiles of a boxplot, the control strategy evaluated can fall into a specific class ranging from 1—representing the highest flexibility and the best future performances—up to 6, representing the worst future performance. For example, for both class 1,2, and 3 the NECV mean is below zero, hence the EFI is positive. Nevertheless, if the control strategy is in class 1, the energy consumption in future scenarios will improve the energy behaviour regardless the scenario considered; while, in class 3, different elements of the matrix could lead to an increase of the energy consumption. Similarly, class 4,5, and 6 represent different distribution of the future results corresponding to negative EFIs: in class 6, the energy performances are always worsened regardless the scenario considered while the number of scenarios with improved energy performances rises moving to class 5 and 4.

Despite the CV threshold guarantees a good representativity of the mean, to further improve the description of the future energy behaviours the authors suggest adding also the CV itself as indicator. Indeed, the higher is the distance of the CV from zero, the higher is the variability of the results in the future matrix. On the contrary, if the CV is zero or close to zero, the EFI is more representative and the behaviour of the control strategy is similar in all the contexts considered in the matrix.

Finally, the brief description of this metric highlights that the use of the EFI allows to summarise several possible future energy behaviours in a single index. Despite the conveniency of having a single indicator, this structure allows to couple this value with a flexibility class and/or the CV to further detail the future behaviour of the technology. The use of these indicators allows to ease the assessment of responsive envelopes—or more in general of a technology or component—in future scenarios providing the tools for a thorough analysis of the energy benefits during

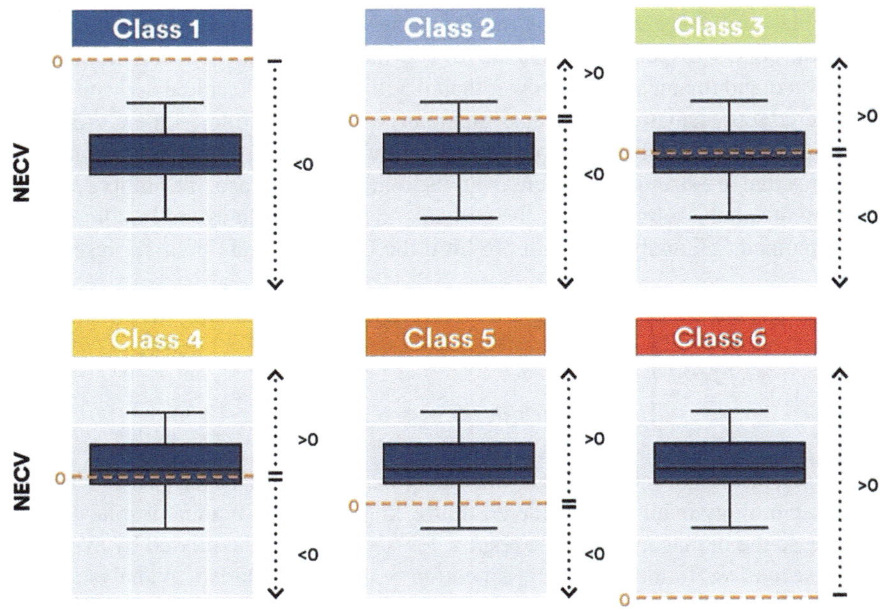

Fig. 6.3 Flexibility classification according to the distribution of the NECV values around the zero [6]

their whole life cycle considering several possible future variations. Therefore, it is highly recommended to implement this metric in the technology assessments considering—where required—specific adaptations to each case study. For example, according to specific analyses, each element of the matrix could have higher or lower probability and, hence, each NECV can be multiplied by a further weighting factor to consider the likelihood of that specific future scenario. Moreover, the same theoretical approach can be easily implemented considering different performance indicators such as, for example, the thermal or daylighting comfort.

References

1. B. Howard, L. Parshall, J. Thompson, S. Hammer, J. Dickinson, V. Modi, Spatial distribution of urban building energy consumption by end use. Energy Build. **45**, 141–151 (2012). https://doi.org/10.1016/J.ENBUILD.2011.10.061
2. V. Lapinskienė, V. Motuzienė, R. Džiugaitė-Tumėnienė, R. Mikučionienė, Impact of internal heat gains on building's energy performance, in *10th International Conference "Environmental Engineering"*, pp. 27–28 (2017). https://doi.org/10.3846/ENVIRO.2017.265
3. S. Karatasou, M. Laskari, M. Santamouris, Determinants of high electricity use and high energy consumption for space and water heating in European social housing: socio-demographic and building characteristics. Energy Build. **170**, 107–114 (2018). https://doi.org/10.1016/J.ENBUILD.2018.04.019

4. A. Nutkiewicz, B. Choi, R.K. Jain, Exploring the influence of urban context on building energy retrofit performance: a hybrid simulation and data-driven approach. Adv. Appl. Energy **3**, 100038 (2021). https://doi.org/10.1016/J.ADAPEN.2021.100038
5. X. Xu, H. AzariJafari, J. Gregory, L. Norford, R. Kirchain, An integrated model for quantifying the impacts of pavement albedo and urban morphology on building energy demand. Energy Build. **211**, 109759 (2020). https://doi.org/10.1016/J.ENBUILD.2020.109759
6. F. Carlucci, K. Negendahl, F. Fiorito, Energy flexibility of building systems in future scenarios: optimization of the control strategy of a dynamic shading system and definition of a new energy flexibility metric. Energy Build. 113056 (2023). https://doi.org/10.1016/J.ENBUILD.2023.113056
7. W. Marion, New typical meterological years and solar radiation data manual (1978). https://doi.org/10.2172/134988
8. W. Marion, K. Urban, *User's Manual for Derived from the 1961–1990 National Solar Radiation Data Base* (1995)
9. S. Wilcox, W. Marion, *Innovation for Our Energy Future Users Manual for TMY3 Data Sets* (2008). http://www.osti.gov/bridge. Accessed 20 Sept 2022
10. H. Kim, K.S. Park, H.Y. Kim, Y.H. Song, Study on variation of internal heat gain in office buildings by chronology. Energies **11**, 1013 (2018). https://doi.org/10.3390/EN11041013
11. R. Andersen, K. Negendahl, Lifespan prediction of existing building typologies. J. Build. Eng. (2022)
12. H. Remøy, T. Van Der Voordt, Adaptive reuse of office buildings into housing: opportunities and risks. Build. Res. Inf. **42**, 381–390 (2014). https://doi.org/10.1080/09613218.2014.865922
13. T.Z. Mine, Adaptive re-use of monuments "restoring religious buildings with different uses." J. Cult. Herit. **14**, S14–S19 (2013). https://doi.org/10.1016/J.CULHER.2012.11.017
14. I.G. Vileniškė, L. Šeduikytė, O. Liutikaitė, Re-functioning of inter-war modernism buildings in Kaunas (Lithuania): the aspect of users' comfort. J. Sustain. Architect. Civil Eng. **20**, 48–57 (2017). https://doi.org/10.5755/J01.SACE.20.3.19323
15. J.M. Kamara, O. Heidrich, V.E. Tafaro, S. Maltese, M.C. Dejaco, F. Re Cecconi, Change factors and the adaptability of buildings. Sustainability **12**, 6585 (2020). https://doi.org/10.3390/SU12166585
16. M. Rinke, R. Pacquée, Structures and change—tracing adaptability based on structural porosity in converted buildings. Archit. Struct. Constr. **1**, 1–12 (2022). https://doi.org/10.1007/S44150-022-00054-9
17. H. Omrany, V. Soebarto, J. Zuo, E. Sharifi, R. Chang, What leads to variations in the results of life-cycle energy assessment? An evidence-based framework for residential buildings. Energy Built Environ. **2**, 392–405 (2021). https://doi.org/10.1016/J.ENBENV.2020.09.005
18. M.T. H.-O. Pörtner, D.C. Roberts, E.S. Poloczanska, K. Mintenbeck, A.O.A. Alegría, M. Craig, S. Langsdorf, S. Löschke, V. Möller, IPCC, 2022: Summary for Policymakers (2022). https://doi.org/10.1017/9781009325844.001

Printed by Printforce, the Netherlands